Process Industries 2

In memory of Jeanne
For Pascal, Christian and Charles
Jean-Pierre Dal Pont

To Annick, Cyril and Jean-Louis
Marie Debacq

Series Editor
Jean-Claude Charpentier

Process Industries 2

Digitalization, a New Key Driver for Industrial Management

Edited by

Jean-Pierre Dal Pont
Marie Debacq

ISTE

WILEY

First published 2020 in Great Britain and the United States by ISTE Ltd and John Wiley & Sons, Inc.

ISTE Ltd
27-37 St George's Road
London SW19 4EU
UK

www.iste.co.uk

John Wiley & Sons, Inc.
111 River Street
Hoboken, NJ 07030
USA

www.wiley.com

Library of Congress Control Number: 2020938699

British Library Cataloguing-in-Publication Data
A CIP record for this book is available from the British Library
ISBN 978-1-78630-562-6

MIX
Paper from responsible sources
FSC® C013604

Contents

Foreword by Laurent Baseilhac

Without doubt, we have entered a new era. The upcoming change seems so significant that there is already talk of a digital revolution. In this context, many questions arise:

– What is the future of the process industry and the future of the talented women and men who are its artisans?

– Will we lose this level of excellence in the field? Or, on the contrary, will we cultivate it by reviving professions and vocations in terms of the challenges we see today?

– Is digital technology a winning bet?

– How are industries preparing for their digital transformation? What are the risks involved?

There are so many thought-provoking new challenges.

Jean-Pierre Dal Pont, Marie Debacq, and their co-authors set out to retrace some industrial trajectories that show that, at different times, industry professionals have had the capacity to overcome the challenges of their time (production, productivity, adaptation to consumption patterns, etc.). The challenges differ today, with societal and environmental markers becoming more and more significant. We understand, through the book, that the answers must probably be sought once again in terms of process engineering technologies, as well as in our ability to renew our management models implemented at the industrial level.

I am sure that readers will share this sense of urgency to tackle all the challenges of the new world. In doing so, we are cultivating a field of expression for our present and future talents and we are working on setting the conditions for a possible reindustrialization.

But let's not go too far, at the risk of betraying the minds of the authors who, at this stage, seek first to provoke the awakening of consciences; they then agree to give us a few solutions, but above all encourage us to pave paths beyond the narrow boundaries that we often draw out of habit in our professions; all this of course without denying the fundamentals that constitute its foundation.

Dear readers, industrialists, academics, students, and future actors or architects of our process industries, I invite you without further delay to dive into this complete, well-documented work, embellished with top quality industrial testimonies where the authors' inextinguishable passion for their industry is evident.

This is a way to no longer doubt the meaning of our profession, a tremendous burst of energy that writes the industrial history of our next decade.

Laurent BASEILHAC
Processes Director, Arkema
Digital Manufacturing Manager

Foreword by Vincent Laflèche

When Jean-Pierre Dal Pont asked me to write a foreword for his book, I accepted it with enthusiasm.

Not just out of friendship. We have known each other and have had the opportunity to collaborate for over 15 years. We share the same belief in the importance of strengthening links and exchanges between industrialists, researchers and teacher-researchers, between private research, academic research, and higher education.

As Deputy Director, then Chief Executive Officer of INERIS (*Institut national de l'environnement industriel et des risques*), then Chairman of BRGM (*Bureau de recherches géologiques et minières*), and, since 2016, Director of the *École des mines de Paris*, I have placed the development of research in partnership with companies at the heart of my strategy, while ensuring that our upstream research activity, funded by public subsidies, constantly feeds the scientific excellence of our teams.

The *École des mines de Paris* adopted its strategic plan in 2017. A member of the new PSL university (*Université Paris sciences et lettres*, which from the outset has been among the top 50 of the best universities in the world and the top French university), the school puts scientists and engineers in an environment as close as possible to a research activity. Over 75% of the school's activity is dedicated to research and only around 25% to teaching. Its ambition is to prepare general engineers capable of making a significant contribution to meet the major challenges of the 21st Century. Ecological transformation, with a particular focus on energy transformation, is clearly a strategic focus. The second area of focus is the digital transformation of companies, with a resolute positioning of project ownership for the school. Since the start of the new school year in September 2019, all our engineering students have common core subjects from the first year, which prepares them for Big Data processing, so as not to use words that sometimes go out of fashion quickly, such as deep learning and artificial intelligence (AI). Ensuring the scientific excellence, in particular mathematical excellence, of our engineering students clearly remains a lasting strategic marker.

The "resolute project ownership" approach means that these tools will subsequently be used in engineering projects, in contact with and paying close attention to industrial partners. The school has a long history of Big Data in geoscience to optimize drilling, whether oil or geothermal. The school was also recognized for the contribution of AI in the detection and treatment of cancers as part of its work with the *Institut Curie*, which is also a member of PSL. The challenge for our graduates is not only to know how to use these new tools, but also to know how to ask and design the industrial question in the wider framework opened up by these new tools, which requires a sound understanding of technological and industrial reality.

We train non-specialized engineers. For more than two centuries, the *mines* engineer has indeed had to integrate scientific, economic, and human dimensions, as well as management, security, openness, and solidarity that the beginnings of a professional career "basically" and inevitably inculcate. For this purpose, the school's training combines the so-called "hard" or engineering sciences, natural sciences, and humanities and social sciences (economics, management, sociology, etc.).

The work of Jean-Pierre Dal Pont and Marie Debacq could not better fit the strategy of the school, and vice versa. The words "theory and practice" have been inscribed on the pediment of our establishment for almost two centuries. Similarly, this work is punctuated and illustrated by concrete cases. This choice can only delight the Director of the *École des mines* that I am. These concrete cases relate to hot topics, often widely publicized (Smart Citites, plastic recycling, etc.). It is not for the sole purpose of following the news: it is recognition of the fact that we are going through a period where we have ever faster cycles of innovation – driven in particular by the digital revolution – and that we need innovation and new technologies to meet the challenges of sustainable development, but also that these innovations are not necessarily accepted by the general public in our society, where we are seeing trust in the engineer and the expert decrease. The engineer and the scientist must integrate this dimension into their approach. The tools and methods developed in this book perfectly integrate these challenges.

This book therefore not only makes an exciting connection between the challenges of business, research, and higher education, it also opens the reader up more broadly to major societal challenges. It does so at a time when, while ecological and energy transitions are underway, the digital revolution is bringing about profound changes to the conduct of companies and industrial management.

This revolution also has consequences on society and brings its share of fears and fantasies, like those engendered by robotics. This book comes at the right time to help students, teachers, researchers, and professionals in their choices and their reflections concerning a rapidly changing world where science and technology are increasingly essential players in sustainable development.

I highly recommend it!

Vincent LAFLÈCHE
Director
École des mines de Paris

Foreword by June C. Wispelwey

Fifteen years ago, when starting the Society for Biological Engineering of the American Institute of Chemical Engineers (AIChE®), I had the good fortune to meet Jean-Pierre Dal Pont. We talked about the future of the chemical engineering profession and about the influence of advancements in biological engineering, particularly with bio-energy and bio-pharma. We discussed the opportunities for creating new life-saving therapeutic proteins and chemicals produced economically from renewable feedstocks. This was the first of many inspiring conversations we would have regarding the future of chemical engineering. It is not a surprise to me that he wrote this passionate book about the process industries and a vision of its future.

Now is a time of transformation. AIChE has a ground floor view of one aspect that is gaining ground – process intensification and modularization. The effort, led by the AIChE's RAPID Manufacturing Institute, is dedicated to improving energy efficiency and lowering investment requirements, and removing barriers that have limited deployment of this technology. For example, process intensification can combine steps and lead to lower costs in industries such as oil and gas, pulp and paper, and chemical production. Modularization enables one to add capacity in small increments which are more suited to a manufacturer's need. The Institute de-risks new technologies in these capital-intensive industries and reduces the ecological footprint.

Another new transformative technology is digitization, or Industry 4.0. There are many aspects of digitization, including the internet of things, smart manufacturing, 3D printing, enhanced or virtual reality, artificial intelligence, big data, robotics, drones and more. These individual technologies are made possible by the new speed of computation, though they are threatened by cybersecurity. Chemical and other engineers, technicians, operators and all those who work in the process industries will need to understand and work with these technologies as they mature.

These volumes arrive at the right time for the new generation, who will enable these technologies and develop new ones to strengthen the process industries and make the world a better place.

June C. WISPELWEY
Executive Director and CEO
AIChE

Introduction

This book, a result of knowledge exchange between the academic and industrial worlds, aims to introduce **process industries** to students, teachers, researchers, professionals, decision-makers, and, in general, the general public, at a time when they are affected by the **digital revolution** that accompanies the ongoing energy and environmental transitions.

These industries aim to transform and/or separate matter by chemical, physical or biological means. They cover huge and often complex fields such as chemistry, petroleum, pharmaceuticals, cosmetics, metallurgy, food industry, biotechnology, environmental and energy industries, among others. Their economic and societal importance is considerable.

The **companies** that depend on it create **value** through their products from industrial facilities (workshops, factories) that implement specific **technologies** and **processes**. The science enabling this implementation is called "chemical engineering" (*génie des procédés* in French).

The French name is to be credited to the late Professor Jacques Villermaux of the *École nationale supérieure des industries chimiques* (ENSIC, the French National School of Chemical Industries) in Nancy, who noted that all the knowledge and techniques of chemical engineering could be perfectly applied, beyond the chemical and petroleum industries, to all process industries.

This book is an invitation to discover the operational modes and technical and industrial management of these industries. It attempts to succinctly answer the following questions:

Introduction written by Jean-Pierre DAL PONT and Marie DEBACQ.

– What is a company?

– What are its foundations and how is it organized?

– How does it respond to what is today known as CSR (corporate social responsibility)?

– How does it cooperate with its stakeholders (clients, stockholders, employees, administration, etc.) when the concept of capitalism with a human face is born which, in addition to remunerating its shareholders, wants to display its contribution to the common good?

– How does it design its commercial products based on the results of its research?

– How does it build and manage its plants and factories to manufacture and distribute its products, after having assessed their impact on the environment through an eco-design analysis based on LCA (Life cycle Assessment)?

– What are the scientific bases and the "technological elements" that the chemical engineer, at the heart of the process, will use to design and operate the **manufacturing facility**?

To ensure their sustainability, process companies must adapt to their socio-economic environment, and, more particularly, to the society they shape through their innovations and products. In particular, they can help respond to the major challenges of today's world, such as that of population growth: if we believe the forecasts, there will be two billion more people to feed by 2050. Growing urbanization will also create quickly insurmountable problems if they are not managed now: a city like Chongqing, on the banks of the Yangtze, has a population that represents half of the population of France. The concepts of Smart Cities and Smart Buildings are therefore essential. As for climate change, this is perhaps the biggest challenge on the planet: the water stress associated with it will affect at least 17 countries, including India. Water is life!

Added to this is the fact that the increasingly enlightened consumer wants to know what they have on their plate, to be informed about the origin of the products they use. Traceability, authentication, naturalness, fair trade, etc. are concepts that manufacturers can no longer ignore. For example, the world is worried about the future of plastics: The Great Pacific Garbage Patch and the North Atlantic Garbage Patch[1], which are several times the size of France, are dumbfounding.

1 Continents of plastic floating on the oceans, sheltering an aquatic fauna that feeds on it and enters the food chain.

This book is particularly interested in the industrial facility at the center of the company. The future of it will depend heavily on its design and its technical and human implementation. Manufacturing operations are no longer considered dirty jobs; it is a given that wealth is built in the workshop (or on the shop floor). Thus, Toyotism, also called "lean manufacturing", is there to prove it: this production system has enabled Toyota to create an empire in the automotive industry and surpass the Americans in their own country.

In recent years, the **digital revolution** has brought about a radical change (disruption) at the societal level and at the level of companies, both at the managerial and productive levels. It was made possible by the increased power of computers (Moore's law), by the multiplication of sensors, their miniaturization, their low cost, and the development of **algorithms**. The notion of artificial intelligence (AI), which brings together a set of **computer applications** and **algorithms** based on the processing and exploitation of Big Data, testifies to this industrial revolution in progress. AI modifies our lives, our professions, our way of moving, very often, of taking care of ourselves, without our being aware of it. This term pervades books, articles, speeches and private and government research programs. Smartphones and tablets, which are only about 10 years old, are one of the essential pieces of media of this revolution. Who could do without it today?

In addition to AI, the digital revolution has brought with it a number of **digital tools** that underpin the concept of the **plant of the future**, born in Germany under the name "factory 4.0". The plant of the future combines the virtual world with the real world. These tools include the IoT (Internet of Things) – everything is connected and everything is connectable – virtual reality, augmented reality, digital twins, additive manufacturing (3D printers), etc. The world of work is deeply affected by robotics and cobotics. We must expect an industry to emerge where repetitive, tiring, messy and even dangerous tasks will be eliminated. The operator will be more of a supervisor than a performer.

Added to this is the fact that the concept of sustainable development, the basis of CSR, is now mature, including the need for metrics. Industry is moving towards a circular, low-carbon and, no doubt, decentralized economy. Bio-industries are not immune to this development with the development of synthetic biology, a remarkable future technological tool, but subject to controversy from the ethical standpoint.

In this shifting context, it is therefore difficult to grasp what the evolution of employment will be; dignified roles are created (Data Officer), while subordinate tasks are on the way out.

Are we moving towards a **civilization of algorithms**? Their opacity raises fears of the advent of a "Black Box Society" where individual freedom is in danger. Everything is known, everything can be known! Our societies – already based on science, technology and knowledge – will become increasingly connected and undoubtedly more complex and more vulnerable.

GAFA (Google, Apple, Facebook, and Amazon), the most powerful digital Internet companies in the world, are already frightening with their capital power, supranationality, and speed of deployment. In this global technological race where everything is accelerating, China has now entered the fray and faces the United States.

These are the reflections that this work invites us to. This book hopes to be interactive and accessible for everyone; it refers to illustrative videos and presents concrete examples, offered by leading figures in the form of boxes. These are listed at the end of each volume.

Videos

The following link to a website makes it easy to access the resources that illustrate this work, in particular, the videos:

https://frama.link/livreIndustriesProcedes

The links and videos are classified by volume and by chapter (via the menu on the left), in the order of appearance in the book.

Volume 1: Sustainability, Managerial and Scientific Fundamentals

Chapter 1: Industries, Businesses and People (Jean-Pierre Dal Pont): this first chapter is devoted to the industry and the businesses that depend on it. It focuses on process industries, while highlighting what differentiates them from the manufacturing and service industries. The themes concerning their constitution, strategy, functioning and governance are discussed.

Chapter 2: Earth, Our Habitat: Products by the Millions, the Need for Awareness (Jean-Pierre Dal Pont and Michel Royer): dedicated to the relationship between products and the environment, this chapter initiates a reflection on our way of life. Earth, our habitat, is a finite space whose complex cycles depend on anthropic activities: we can cite, for example, atmospheric chemistry and the problem of ozone. The vital systems of water, food, energy and climate are referred to as a "nexus", because they are interdependent. Products, whose quantity is increasing with the population explosion, must be ecodesigned using LCA (Lifecycle Assessment), toxicology, ecotoxicology and traceability studies, and turn to biobased raw materials. The circular economy must prevail over a linear economy, which consists of extracting, producing, consuming and throwing away.

Chapter 3: Designing Chemical Products (Willi Meier): Chapter 3 is dedicated to product design and formulation. A product must be designed to meet the needs of customers. In now saturated markets, companies are turning to often complex functionalized products. Post-its are a vivid example: at first, it was just a glue that stuck badly! Who could do without them today? Increasingly based on bio-sourced raw materials and biotechnologies, products use additives: ingredients such as starch and gelatin. This is the case for drugs that can also be encapsulated with alginates to reach the right target at the right time. The story of Aspirin®, first synthesized by Bayer in 1897, is remarkable. Its survival is due, in part, to sophisticated formulations. Another example of the development of drinkable formulations is coffee. The formulation of environment-friendly "smart" products in the field of textiles or fertilizers, for example, is a science with a bright future.

Chapter 4: Chemical Engineering: Introduction and Fundamentals (Marie Debacq, Alain Gaunand and Céline Houriez): chemical engineering, although omnipresent, is almost unknown to the general public. The beginning of this chapter therefore endeavors to give some definitions and historical benchmarks about this young applied science. The fundamentals of chemical engineering are then presented: starting with thermodynamics, then transfers, and finally chemical kinetics and catalysis. The last part of the chapter presents the "system-balances-performance" approach for process design using two simple examples. A box presents the very first level of calculation on processes, namely material balances.

Chapter 5: Chemical Engineering: Unit Operations (Marie Debacq): the concept of a unit operation has made it possible to bring together, in large categories, the innumerable equipment used by the process industries. There are numerous unit operations and there is no a question of giving an exhaustive presentation here. This chapter therefore covers some of them, chosen because they are particularly symbolic or representative of one type of operation or another. Thus, the following are presented: distillation, the most important separation operation and also certainly the most scientifically mature; some fluid/solid mechanical separation operations, very widespread industrially but still relatively empirical today; agitation, as a symbol of the importance of hydrodynamics (that is to say, the study of fluid movements) in chemical engineering; heat exchangers, the main representatives of transfer operations (heat exchangers dealing with the process of heat transfer); and, finally, reactors, which are at the heart of processes and responsible for the transformation of matter on the scale of the molecules themselves.

Volume 2: Industrial Management and the Digital Revolution

Chapter 1: Bio-industry in the Age of the Transition to Digital Technology: Significance and Recent Advances (Philippe Jacques): the digital revolution is profoundly changing the profession of engineers involved in bio-industries. This chapter describes the main stages of development of a product of microbial origin and how approaches related to bioinformatics, synthetic biology, systems biology and microfluidics will make it possible to amplify the development of this growing economic sector.

Chapter 2: Hydrogen Production by Steam Reforming (Marie Basin, Diana Tudorache, Matthieu Flin, Raphaël Faure and Philippe Arpentinier): this chapter presents the most widely used hydrogen production process in the world: steam reforming of natural gas. All the technological elements of this process are described, as are the problems of industrial operation of these units. Current and future developments, including those aimed at minimizing carbon dioxide emissions, are also discussed.

Chapter 3: Industrialization: From Research to Final Product (Jean-Pierre Dal Pont): the process includes all the technologies that plants and factories use to manufacture a product or a set of products. Very generally, this is a reaction followed by purification: a drug or a product to protect plants, often complex molecules, are the result of several reactions and several separations or purifications called "unit operations", described elsewhere.

The purpose of this chapter is to describe the industrialization process, which, starting from research, will define the production tool. At the end of the chapter, two boxes describe the increasingly sought-after modular construction and the constraints and advantages of a multi-workshop platform.

Chapter 4: Operations (Jean-Pierre Dal Pont): operations, or manufacturing, designate the implementation of industrial facilities (plants or factories). They are an essential function of the process industries, the source of their products and related services, and, therefore, of their profit.

This chapter studies production, its flows (financial, information, materials), and the increasingly sophisticated IT tools that make it possible to manage them such as ERP (Enterprise Resource Planning). It also discusses the bases for calculating the cost price of products and margins. Finally, special thought is given to change management: to last is also to change.

Chapter 5: The Enterprise and The Plant of the Future at the Age of the Transition to Digital Technology (Jean-Pierre Dal Pont): Chapter 5 recalls the industrial revolutions that have followed one another since the invention of the steam engine, a source of energy at the beginning of the 18th Century, to the present day. It analyzes their impact on society and on the capital-intensive business as we know it today. Emphasis is placed on information technology, which took off after the Second World War. The emergence of the Internet around 1990, that of the smartphone around 2000, and the beginnings of artificial intelligence initiated the digital revolution, whose unprecedented impact we are already seeing on society and industry. Many boxes give examples of the use of AI in fields as varied as autonomous cars, underwater exploration, robotics and industrial management.

Chapter 6: And Tomorrow... (Jean-Pierre Dal Pont): this last chapter is a reflection on the digital revolution as it is perceived today and, more particularly, on artificial intelligence, which is its standard-bearing media. AI is increasingly affecting the city which wants to be *smart*. The water sector is taken as an example of economic activity whose digital aspect modifies the processes, the management of the distribution networks and the trades.

While the various applications of this emerging technology can hold out hope for many advances and improvements, the use of AI raises many questions. The very functioning of the industrial business is turned upside down. Will Teslism, synonymous with, among other things, the "hybridization" of computer systems, supplant Fordism? Isn't the robot assisting the operator a threat to his job? The citizen, meanwhile, questions the intrusion of GAFA in his private life and governments about their supranationality. The "fully connected" raises fears for the fragility of administrative and industrial systems, while cybercrime is a ubiquitous threat. The fundamental question is whether human beings are at the heart of the system and... for how long.

In the current period of upheaval where "the only certainty is uncertainty", perhaps we must take one of the thoughts of the great manager of the 20th Century, Peter Drucker: "The best way to predict the future is to create it." One of the ambitions of this book is to help the readership in this research, or at least, to try to whet their curiosity.

1

Bio-industry in the Age of the Transition to Digital Technology: Significance and Recent Advances

1.1. Introduction

Bio-industry involves living organisms, mainly microorganisms, bacteria*[1] and fungi*, and develops products for the agrifood industry (yeasts*, lactic ferments*, enzymes*, etc.), the pharmaceutical industry (vaccines, antibiotics, etc.), agriculture (biopesticides*, biostimulants*, etc.), the environment (microorganisms for water, air and soil depollution, etc.), chemical industries (synthons*, biodegradable polymers, biosurfactants*, etc.). The markets concerned by these products are steadily increasing by several percent each year. For example, fine chemicals originating from fermentation were estimated at $24 billion in 2017, with a yearly increase of 3.4% over the 2017–2022 period (source: BCC).

The beginnings of bio-industry are probably in brewing beer in Mesopotamia, about 6,000 years ago, and wine production in Egypt as early as the Middle Predynastic period, more than 5,000 years ago (Figure 1.1). Over thousands of years, microorganisms have been used to produce (alcoholic beverages) or preserve (fermented products) numerous foods and have been used in processes such as flax retting (an operation that involves exploiting the production capabilities of hydrolytic enzymes in soil microorganisms to isolate the fibers of flax stems) without the people utilizing them even noticing their existence.

Chapter written by Philippe JACQUES.

1 The asterisk symbols refer to the glossary at the end of the chapter.

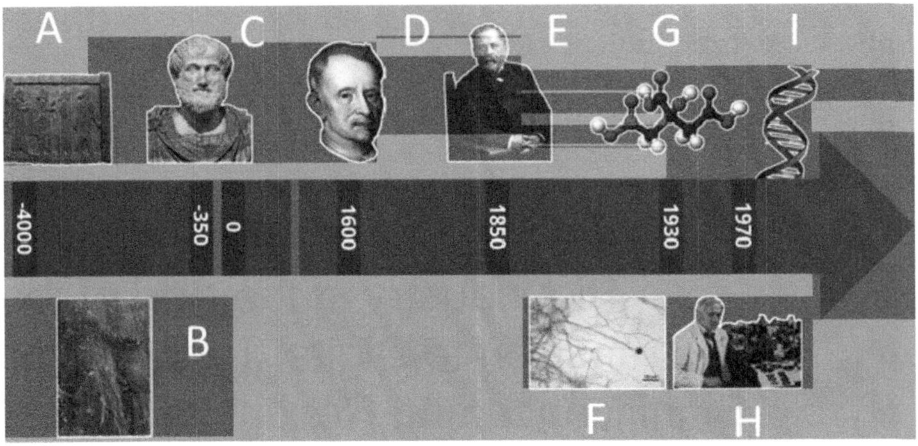

Figure 1.1. *Evolution of industrial microbiology. For a color version of this figure, see www.iste.co.uk/dalpont/process2.zip*

COMMENT ON FIGURE 1.1.– *The beginnings of industrial microbiology probably date back to the brewing of the first beers in Mesopotamia, some 6,000 years ago (A). During millennia, microorganisms have been exploited without knowledge of their existence, for example in flax retting (B). The spontaneous generation theory devised from Aristotle (C) to Van Helmont (D) has even been a major impediment to the demonstration of this form life invisible to the naked eye. This was definitively demonstrated due to the works of Antoni Van Leeuwenhoek and Louis Pasteur (E) during the 18th and 19th Centuries. During the first part of the 20th Century, various industrial applications involving microorganisms were created, such as for the production of amylase, citric acid (G) or penicillin, discovered by Fleming (H), using fungi (F). The discovery of DNA (I) in 1953 opened the field of genetic manipulation of microorganisms.*

This period precedes the works of Antoni Van Leeuwenhoek (1632–1723) and Pasteur (1822–1895). The first, through the invention of a tool, the microscope, capable of visualizing microscopic living beings, was able to highlight the presence of these microorganisms in many environments. The second has definitively demonstrated their role in the processes of processing organic matter and thus put an end to the theory of spontaneous generation (appearance of living beings from non-living matter) defended for centuries by renowned scientists.

Subsequently, researchers, now aware of the existence of life at the microscopic scale, were able to consider diversifying its exploitation for the production of organic acids, enzymes or antibiotics such as penicillin. The latter was discovered by chance in 1928 by a Scottish physician and bacteriologist, Sir Alexander Fleming. It was, in fact, following contamination of his samples by a fungus called *Penicillium notatum* that he was able to identify a substance capable of inhibiting the growth of pathogenic bacteria under laboratory conditions. Fleming's work was pursued by an Australian and a German, Florey and Chain in 1939, who demonstrated the potential of penicillin to fight bacterial infections in animals. Penicillin was then produced in industrial quantities during World War II and thus saved thousands of lives on the battlefield. Fleming, Florey and Chain received the Nobel Prize for this discovery in 1945.

The discovery of deoxyribonucleic acid* (DNA) by Watson and Crick in 1953 and the development of molecular biology techniques have allowed the manipulation of the genetic heritage of microorganisms to increase the synthesis potential of enzymes and vaccines, among others. These technologies are probably reaching their peak today with the development of **synthetic biology**. The latter aims to design new microorganisms by modifying more or less profoundly the metabolic pathways of a natural microorganism. These microorganisms, which are true cell factories, should eventually be optimized for the production of many molecules of interest under perfectly defined conditions.

With the development of digital technology, a new era is opening up for industrial microbiology: that of **bioinformatics** applied to microorganisms and **systems biology**. The latter approach aims to develop a global view of how known microorganisms function by modeling, among other things, the sequence of multiple intracellular biochemical reactions, as well as their mode of regulation, in order to optimize their functioning in a given direction. Its aim is to manage and integrate the large amounts of data accumulated today, in particular as a result of the development of high-throughput approaches in so-called "omics": **genomics***, **proteomics***, **metabolomics***, **transcriptomics***, etc. To these approaches, one should also add the possibility offered by the potential of **microfluidics*** to analyze and control the behavior of each cell within a population, but also to revolutionize strategies for increasing the production volume by shifting from the notion of scaling-up to a concept of scale reduction.

Equipped with these new tools, **modern industrial microbiology** nowadays faces enormous challenges. High-throughput screening approaches to the potential for synthesis of microorganisms, of which only a tiny fraction is understood today, must strive to meet an ever-increasing demand for more biodegradable, less toxic products and whose production is based on eco-processes. These bioproducts should eventually replace a significant part of the synthetic molecules that, as a result of the development of chemistry, have invaded our environment. In addition to identifying suitable substitutes, engineers will therefore need to develop microbial strains* as well as bioprocesses* whose productivity will contribute to the achievement of cost prices close to the prices of the synthetic molecules to compete with. At the moment of writing, this last challenge remains the major obstacle to the breakthrough of a number of products of microbial origin.

1.2. Diversity of products and applications

The microorganisms used today in the industrial world belong to the domain of organisms known as prokaryotic or eukaryotic. The former is characterized by an absence of nucleus but also an intracellular organization poorly structured in organelles. These are bacteria and archaebacteria*. The latter include microorganisms whose cells contain a nucleus and a large number of various organelles. The main industrial eukaryotic microorganisms are yeasts, molds* and microalgae* (Figure 1.2).

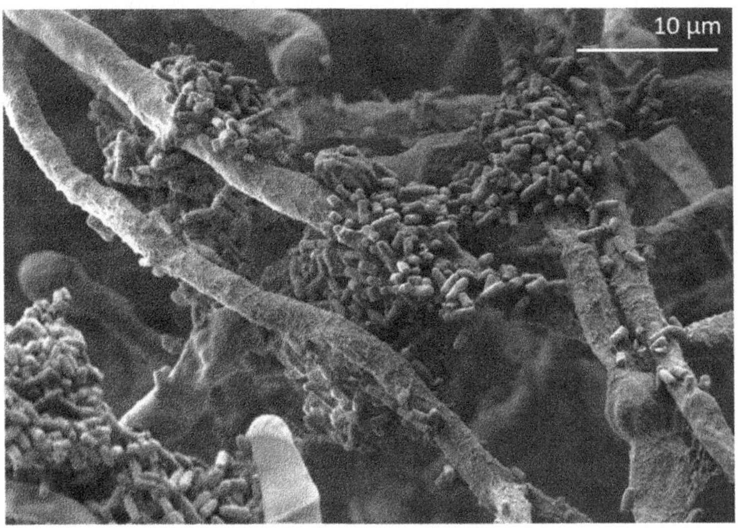

Figure 1.2. *Electron microscopy analysis of a bacterium and mold coculture. The bacterium is represented by the small sticks and the mold by the long filaments*

The applications of industrial microbiology cover a very large number of sectors of activities. At the European level, a color classification of biotechnology application sectors, including industrial microbiology, was proposed a few years ago (Figure 1.3). **Red biotechnology** concerns the healthcare sector; **green biotechnology** is linked to agriculture and agri-food sectors; **yellow biotechnology** includes applications related to the protection of the environment, to pollution treatment or elimination; **white biotechnology** develops bioprocesses and substitutes for the chemical industry; finally, **blue biotechnology** studies products related to marine biodiversity. The evolution of the biotechnology market greatly varies depending on the sector concerned. More than 50% of new pharmaceutical products stem from red biotechnology. The development of **biopesticides** as a substitute for pesticides is an example of the development of green biotechnology. While these biopesticides currently only account for between 5% and 10% of synthetic pesticides, their market is growing by about 14% per year, while the global market for synthetic pesticides has reached a plateau in recent years. In the field of white biotechnology, enzyme* production is a booming market for textile, leather, biofuel, detergence, etc. industries. The other flagship of white biotechnology is the production of bioethanol. The latter must now evolve to make use of the processing of crop residues rather than compete with food-oriented crops. Yellow biotechnology, as well as green biotechnology for that matter, are facing cost prices too high compared to concurrent chemicals and restrictions on the European market related to the use of genetically manipulated microorganisms. Finally, the exploitation of the marine microbial potential still remains in its infancy.

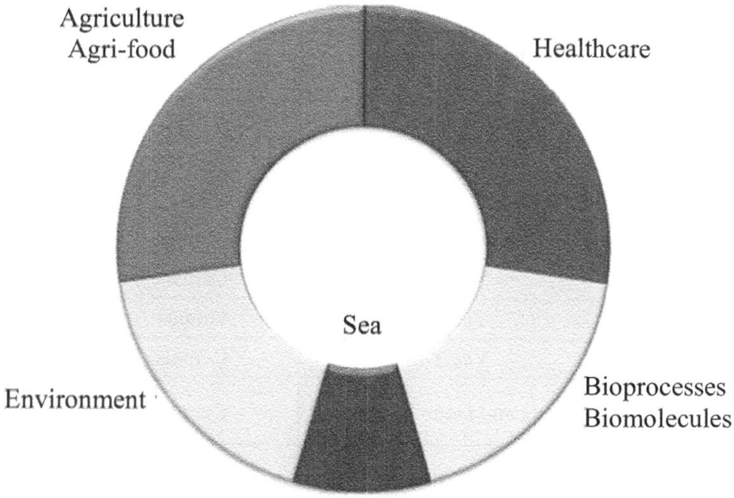

Figure 1.3. *Classification of biotechnologies according to a color code. For a color version of this figure, see www.iste.co.uk/dalpont/process2.zip*

1.2.1. *Fermentations in agri-food*

Industrial microbiology finds its origins in exploiting microorganisms for the processing and storage of products of plant or animal origins. Indeed, many microorganisms convert carbohydrates into organic acids (lactic fermentation for example) or alcohol (alcoholic fermentation). Other bacteria are also capable of developing what is known as alkaline fermentation, which aims to produce ammonia from the degradation of protein sources. The various bacteria and yeasts involved in these processes have for many centuries been exploited through the use of part of the fermented product in which they are contained to inoculate a fresh product and repeat the transformation phenomenon. The needs of process standardization have led companies to use microorganisms themselves produced in dedicated companies to obtain reproducible fermentations that meet the quality criteria of the food sector. These products are now known as **starter cultures**, and they are living microorganisms sold in liquid or dry form and capable of starting a fermentation process. The quintessential example of this development is baking yeast (*Saccharomyces cerevisiae*) which is today the product based on the world's most produced living biomass of microorganisms. Other starter cultures are used in agri-food, such as lactic bacteria which ensure the fermentation of lactose of milk in lactic acid and constitute the basis of the production of numerous dairy products such as yogurts or cheeses, but also meat fermentation in sausage or that of cabbage for sauerkraut. Table 1.1 includes some examples of fermented products developed around the world.

Product name	**Microorganisms involved**	**Agro-resource**	**Type of fermentation**
Sauerkraut	Lactic bacteria	Cabbage	Lactic
Sausage	Lactic bacteria	Meat	Lactic
Yogurt	Lactic bacteria	Milk	Lactic
Cheese	Lactic bacteria	Milk	Lactic
Beer	Yeast	Cereals	Alcoholic
Wine	Yeast	Grapes	Alcoholic
Bread	Yeast	Cereals	Alcoholic
Sourdough bread	Yeasts and lactic bacteria	Cereals	Alcoholic and lactic
Kefir	Yeasts and lactic bacteria	Milk	Alcoholic and lactic
Miso	Mushroom and lactic bacteria	Cereals	Lactic
Netetu	*Bacillus*	Parkia biglobosa seeds	Alkaline
Natto	*Bacillus*	Soy	Alkaline

Table 1.1. *Examples of fermented products*

1.2.2. *Biomass-based products*

The starter culture concept has gradually expanded to other applications, such as enzyme-producing microorganisms whose hydrolytic properties are exploitable in environments such as septic tanks, grease traps used in water purification processes or for bioremediation* of polluted soils.

Another example is that of microorganisms naturally present in so-called disease suppressing soils and therefore potentially capable of performing a controlling activity of predators in certain crops (biopesticides).

Table 1.2 includes some examples of such products.

Name	Property	Microorganism
Lactic ferments	Sugar transformation into organic acids	*Lactobacillus, Lactococcus,* etc.
Probiotics	Easier food digestion, protection from pathogens	*Bifidobacterium*
Decontamination starters	Degradation of xenobiotic (polluting) substances	*Pseudomonas, Rhodococcus,* etc.
Degradation starters	Fat degradation	*Bacillus, Yarrowia lipolytica*
Biopesticide	Antagonistic activity against a plant crop predator	*Bacillus thuringiensis, Bacillus velezensis, Pseudomonas fluorescens, Beauveria*
Alcoholic ferments	Sugar transformation into alcohol and carbon dioxide	*Saccharomyces cerevisiae, Saccharomyces carlsbergiensis*

Table 1.2. *Examples of starter cultures*

1.2.3. *Metabolite* based products*

The other large family of products of microbiological origin is that of biomolecules produced by microorganisms. Their very important biodiversity ranges from simple organic acids created by the central metabolism to more or less complex polymers that act as energy storage for the cell, including the panoply of proteins or enzymes or even metabolites secondary to multiple structures and applications. An enzyme, amylase, capable of degrading starch, was one of the first metabolites produced on an industrial scale from the *Aspergillus niger* mold.

Genetic engineering techniques have also produced metabolites of plant or animal origin from genetically manipulated microorganisms. The first example is insulin, an animal hormone produced by a processed bacterium, *Escherichia coli.*

Nowadays, enzymes, metabolites and polymers of microbial origin are extremely varied. Table 1.3 takes a non-exhaustive view of their diversity by specifying their current market.

1.3. Traditional process for developing a product of industrial microbiology

The development of a product for industrial microbiology traditionally involves the following steps:

– **a selection stage** of a microorganism that meets a number of criteria. This organism often comes from a natural environment in which it exercises a property of interest;

– an **improvement** of the properties of the microorganism selected by a random mutagenesis or genetic manipulation approach;

– the development of a **small-scale production** and **purification** process;

– the large-scale **transfer** of this process;

– the product **formulation**.

For many decades, this strategy has been used with fairly similar tools regardless of the product or application under consideration. Today, technological developments are profoundly changing this strategy. In the rest of this chapter, these developments will be presented during the various stages of the process. Combined, they most likely constitute the typical blueprint **for the futuristic factory** in industrial microbiology.

Ingredient type	Global market 2017 ($ millions)	Main products	Market by product ($ million)	Main application sector
Organic acids	3119	Citric acid	1714	Food and beverage, Pharmaceutical and cosmetics, Feed, Industrial and others
		Gluconic acid	232	Food and beverage, Pharmaceutical and cosmetics, Industrial and others
		Itaconic acid	115	Industrial and others
		Lactic acid	1010	Food and beverage, Pharmaceutical and cosmetics, Industrial and others
		Succinic acid	48	Food and beverage, Pharmaceutical and cosmetics, Industrial and others
Amino acids	9271	Glutamic acid	3678	Food and beverage
		Lysine	4410	Feed, Industrial and others
		Threonine	1107	Pharmaceutical and cosmetics, Feed
		Tryptophane	76	Pharmaceutical and cosmetics, Feed
Vitamins and carotenoids	1981	Ascorbic acid	954	Food and beverage, Pharmaceutical and cosmetics, Feed, Industrial and others
		Iso ascorbic acid	125	Food and beverage, Pharmaceutical and cosmetics, Industrial and others
		Carotenoids	270	Food and beverage, Pharmaceutical and cosmetics
		Riboflavin	504	Food and beverage, Pharmaceutical and cosmetics, Feed
Enzymes	6287	Carbohydrases	1791.2	Food and beverage, Pharmaceutical and cosmetics, Feed, Industrial and others
		Proteases	1344.4	Food and beverage, Pharmaceutical and cosmetics, Feed, Industrial and others
		Polymerases and nucleases	932.4	Pharmaceutical and cosmetics, Industrial and others
		Lipases	244.3	Food and beverage, Pharmaceutical and cosmetics, Feed, Industrial and others
		Phytases	259.9	Feed
		Other enzymes	681.9	Food and beverage, Pharmaceutical and cosmetics, Feed, Industrial and others
Polysaccharides and polymers	769	Xanthan	739	Food and beverage, Pharmaceutical and cosmetics
Antibiotics	2583	Penicillin	756	Pharmaceutical and cosmetics, Feed

Table 1.3. *Market for metabolites of microbial origin*

1.4. Strain selection and optimization

1.4.1. *Evolution of strain screening techniques*

It is very difficult to quantify the exploitable potential of microbiological life known today, but it is clear that it constitutes a tiny part of what remains to be discovered. The selection of microorganisms exhibiting certain properties thus remains one of the objectives of a large number of companies. Knowing that one gram or milliliter of organic medium can contain billions or tens of billions of microbial cells, selecting the best performing microorganism is often a laborious task which proves particularly labor intensive. However, this approach is now benefiting from advances in **robotization* and device miniaturization***.

In the past, the selection of microorganisms was primarily considered on solid medium. The objective was to develop a technique making it possible to identify a specific phenotype in a colony among tens or hundreds on a Petri* dish with an agar culture medium adapted to selection. The minimalist standard was the selection of the best strain from the analysis of 10,000 clones. The properties sought are very diverse: selection of lactic bacteria with a high acidification power by way of bringing forward a solubilization of a substrate or the evolution of pH by means of an indicator; identification of the hydrolytic properties (proteases on a casein-based medium, amylase on a starch-based medium, etc.) of antagonistic properties (inhibition of the growth of a bacterium, a yeast or a fungus (Figure 1.4)) or biosurfactant properties (colony size on an medium whose agar concentration is reduced).

The use of a robotic tool such as the **colony picker** (Figure 1.5) has allowed the automation of this approach and, speeding up the selection process or multiplying the number of clones analyzed. Using a camera, this tool is capable of analyzing changes in turbidity or the color of the environment of a colony, or even colony sizes, and is able to transfer the most efficient clones into a liquid medium for further testing or conservation. Depending on the equipment, this task can be performed at a speed of a few hundred to several thousand transplanted clones per hour.

Nonetheless, the evolution of robotization and miniaturization technologies of reactors also makes it possible today to consider the selection of microorganisms of interest on the basis of their behavior in liquid environments. The possibility of cultivating strains in microtiter plates ranging from 24 to 396 wells (**microreactors**) from a few milliliters to a few dozen microliters allows one to consider testing a large number of clones in conditions that are often closer to those that will then be used for large-scale production.

These technologies also offer a larger selection test panel. Robotization now allows, among other things, to collect, add, mix, incubate, filter and automatically detect. The result is a multitude of possibilities for analyzing the potential of the strains to be selected (Figure 1.6).

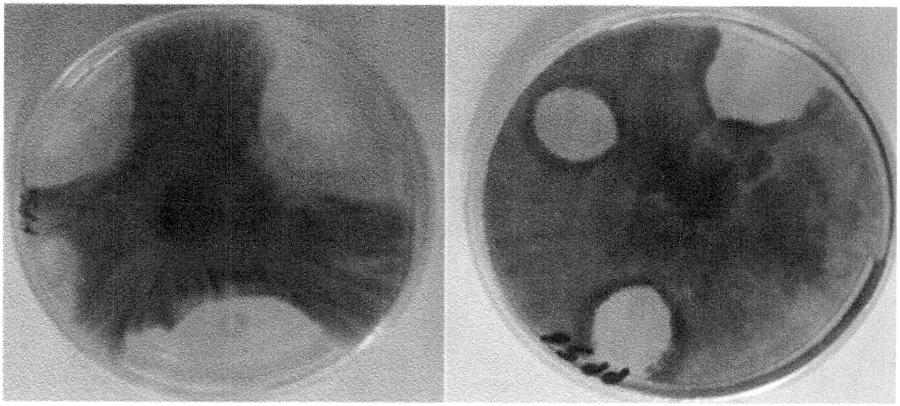

Figure 1.4. *Examples of selection techniques on solid medium. Research of antagonistic bacteria of the growth of a phytopathogenic fungus (*Fusarium oxysporum*) for the development of new biopesticides. For a color version of this figure, see www.iste.co.uk/dalpont/process2.zip*

Figure 1.5. *The colony picker is a piece of equipment that enables high-throughput screening of microbial strains possessing a particular property. It works by means of a camera that tracks colonies of microorganisms presenting a different phenotypic character (color, shape, pigment production, etc.)*

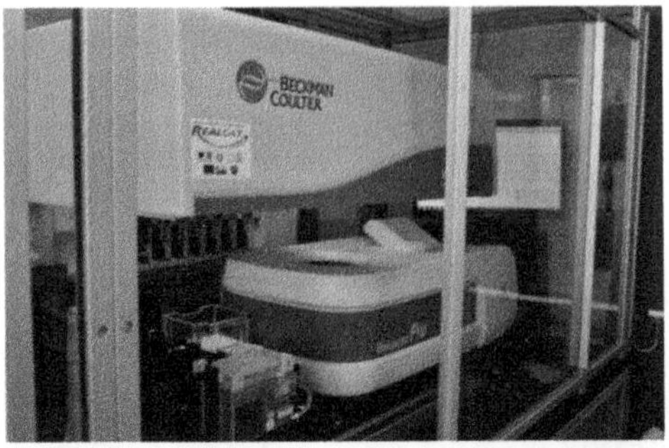

a)

b)

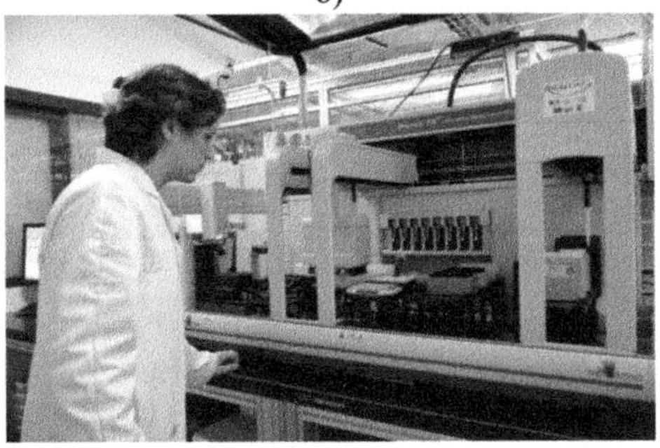

c)

Figure 1.6. *Examples of robots for microorganism culturing and the preparation of culture media*

COMMENT ON FIGURE 1.6.– *This type of equipment makes it possible to automatically prepare with a high-throughput culture media for reactors from a few hundreds of microliters to a few milliliters (c). These environments are then incubated and stirred in a device such as the "biolector" which allows for online monitoring of pH, dissolved oxygen, biomass growth and the concentration of a fluorescent probe ((a) and (b)).*

This device still naturally presents a number of limitations. For example, the potential of microbiological life in the seabed is nowadays practically unknown. Revealing it with robotic tools requires highly pressurized reactors. These devices are not yet available for high-throughput screening.

The explosion of genome sequencing of microorganisms and the improvement of bioinformatics prediction tools are now making it possible to consider new high-throughput screening* strategies. It is thus possible to search for a gene encoding a particular enzyme, allowing the expression of a particular phenotype* or a cluster of genes* encoding the enzymes responsible for the synthesis of a metabolite presenting a structure or particular structural features. This selection is thus achieved firstly *in silico**, that is to say without experimenting in a laboratory. It should enable saving valuable time in years to come.

1.4.2. *Evolution of genetic modification technologies, from random mutagenesis* to CRISPR-Cas9 technology*

The optimization of microbial strains has, for many years, been based on the use of random mutagenesis* obtained through cell treatment by irradiation (ultra-violet, X-rays, etc.) or by the use of chemical mutagenic agents (N-methyl-N'-nitro-nitrosoguanidine (NTG), ethyl methanesulfonate (EMS), etc.). Overproducing penicillin strains obtained after successive treatments of *Penicillium chrysogenum* by these different physical or chemical agents and the selection of hyperproducing mutants is a well-known example thereof. It is estimated that some twenty successive mutation stages have led to an increase from an antibiotic production of 60 mg/l to 7 g/l which corresponds to a productivity 100 times higher.

The advent of genetic engineering techniques and the in-depth knowledge of metabolic pathways of metabolite* biosynthesis of interest have gradually enabled a shift towards the targeted strain optimization by modifying the expression of certain genes (overexpression of an enzyme by using strong promoters*, for example) or by interrupting the latter. These approaches have led to metabolic engineering, which consists of directing the metabolism of a given strain in order to make it produce the maximum amount of metabolite under consideration. The latter makes use of the evaluation of metabolic flows that characterizes in moles/kg of dry matter and per hour the specific transformation rate of each reaction substrate in its product.

Three dimensions have recently revolutionized these approaches. On the one hand, advances in genetic techniques are now allowing genetic modifications to be multiplied while eliminating the selection markers used at the end of the operation (auxotrophy*, antibiotic resistance, etc.). Technologies such as **CRISPR-Cas9** (Clustered Regular Interspaced Short Palindromic Repeats) also make it possible to limit genetic modifications to their strict minimum. These tools, which allow for minimal genome limitation without the addition of genes or foreign DNA to the strain, could ultimately constitute a response to requests for a reduced use of GMOs. A second dimension is the biology of systems, which consists of encompassing and modelling as many phenomena as possible occurring within a cell or a population of cells to optimize their functioning as part of a given objective. These models allow, with a number of strains well-controlled at the industrial scale such as *Escherichia coli*, *Bacillus subtilis* or bacteria or yeasts such as *Saccharomyces cerevisiae*, *Pichia pastoris* or *Yarrowia lipolytica*, to provide a global view of their metabolism or the regulation thereof and to define targeted modification strategies of their genetic heritage. A third approach is that of synthetic biology* which aims to induce the production by a given microorganism of a metabolite that it was unable to synthesize by transferring into this microorganism the genes encoding the enzymes responsible for the entire metabolic pathway necessary for this synthesis. Another approach to synthetic biology is one that consists in defining the minimum genome necessary to the growth of a microorganism in a given environment by eliminating all non-essential genes to cell life. This approach should eventually define "microbial chassis*" ready to synthesize metabolites of different origins. These microbial chassis could be adapted to a particular inexpensive substrate.

1.5. Production and purification processes

1.5.1. *Needs of microorganisms*

The growth of a microorganism and/or the production of a metabolite of interest depend heavily on the environment in which the microorganism is located. The first need therefore is to have access to an energy source. For a certain number of microorganisms exploited in industrial microbiology, this energy is drawn from the oxidation of an organic substrate such as glucose throughout a process called respiration*: they are called chemoorganotrophic* organisms. For so-called aerobic microorganisms, the final electron acceptor is oxygen. The latter is not very water-soluble (7.55 mg/l at 30°C and 1 bar), it is the transfer rate from a gas phase to a liquid phase that often constitutes a limiting factor. For optional anaerobic microorganisms (namely those that are able to develop in the absence and presence of oxygen), the electron acceptor is an organic molecule. Finally, strict anaerobic microorganisms use a mineral molecule (nitrate, sulphate, etc.) as an electron acceptor. Chemolithotrophic* organisms draw their energy from the oxidation of a mineral substance. Finally, the third major source of energy used is light for photo-organotrophic* organisms.

The second essential element is to be able to extract from the culture medium all the elements (C, H, O, N, P and S) necessary to manufacture the set of monomers (amino acids, monosaccaride, fatty acids, puric and pyrimidic bases, vitamins, etc.) and polymers (proteins, polysaccharides, lipids, nucleic acids, etc.) that compose a microorganism. In addition, there is a certain number of trace elements necessary to ensure ion and osmotic equilibrium in the cell but also the realization of a number of reactions that require their presence as a cofactor. They are often classified according to the needs of the cell, in major (Na, K, Cl, Mg, Mn, etc.) and minor (Fe, Cu, Co, etc.) trace elements. Some microorganisms may merely require a source of organic carbon and different mineral salts containing nitrogen, phosphorus, sulfur and trace elements to develop. Others require a much richer medium in organic molecules that they have lost the ability to manufacture on their own. Photosynthetic microorganisms such as microalgae are able to use CO_2 as a carbon source. Finally, every strain of microorganism develops within a pH and temperature range that is specific to it. This mix of different information conditions the environment in which microorganisms are able to develop optimally and thus the processes developed to produce them.

1.5.2. *Production processes*

Depending on the nature of the microorganism produced and its application, production processes will have to meet all or part of the following criteria:

– allow a pure microorganism culture or a controlled culture of a microbial consortium;

– ensure a controlled supply of oxygen or CO_2 in gas form;

– allow pH control;

– allow temperature control;

– allow continuous substrate feeding;

– allow the control of foam formation;

– ensure adequate contact with a light source (for microalgae cultures);

– ensure a homogeneous mixture of a system containing three phases: a solid phase (microorganisms), a liquid phase (the culture medium) and a gas phase (compressed air, oxygen or carbon dioxide).

Inherited from chemical engineering, the traditional bioreactor used today in the vast majority of bio-industries meets the majority of these criteria. In short, its mirror-polished stainless-steel structure allows it to avoid the adhesion of microorganisms on the reactor walls while ensuring resistance to the steam sterilization process of the equipment. Equipped with a heat exchanger (usually a double envelope) to maintain the temperature, it possesses a gas supply system (air, oxygen, CO_2, etc.) most often equipped with a sterilizing filter and a mixing system that ensures a homogeneity of the three phases existing within the bioreactor. This system also plays an important role in optimizing the transfer of oxygen or CO_2 from gas bubbles to the culture medium. The reactor also includes a sample collection valve, a drain valve, as well as one or two valves for adding liquid. An air outlet often equipped with a sterilizing filter ensures the evacuation of gases. A number of sensors allow the online monitoring of certain parameters such as pH, temperature, dissolved oxygen concentration, volume, presence of foam, etc. Some of these parameters can be regulated. These sensors are therefore connected to an automat which, under the control of supervising software, sends messages to actuators, which are most often valves.

Since microalgae culturing requires a significant amount of light, it employs very different types of reactors. These latter must ensure a maximum transfer of photons and are therefore often based on maximal exposure (with a large surface-to-volume ratio) of culture to light. In addition, the absence of carbon source in the medium reduces the risk of contamination and therefore makes it possible for some productions of open-pit cultures to be considered.

Three traditional cultivation techniques (Figure 1.7) are being considered in industrial microbiology. In the so-called "batch" culture, a bioreactor containing all the elements necessary for the growth of a microorganism and/or the production of a metabolite is seeded by a preculture (inoculum). No additional culture medium is added or extracted during production. This simple process is often the one used in a first production approach. In continuous cultivation, fresh culture medium is continuously brought and the fermentation medium is also extracted with an identical flow to ensure a constant culture volume. The dilution rate used imposes the growth rate of the microorganism if it is lower (chemostat-type process) or equal (turbidostat-type process) to the maximum growth rate of the microorganism. This process is relatively unused at the industrial level. In addition to technical and logistical difficulties (maintaining the sterility of a complex process and the need for continuous monitoring), it poses a significant risk for certain strains of genetic drift following the appearance of spontaneous mutants potentially better adapted to the culture mode. Rather, it is used at the laboratory level to better understand the physiological behavior of a given microbial strain. The third process, "fed-batch", is therefore the one traditionally employed by manufacturers to ensure increased productivity compared to batch culture. It consists of ensuring a gradual feeding of the culture medium without removing culture broth. It can be used to avoid too high a concentration of substrate in the culture medium, which would directly or indirectly cause a potential inhibition of the microorganism growth (such as in the production of bakery yeast). The other application of this device is the possibility of ensuring an optimal growth in batch conditions for a given microorganism and in order to then bring it under different culturing conditions, by adding a particular medium, to optimally produce a metabolite of interest (such as for the production of penicillin).

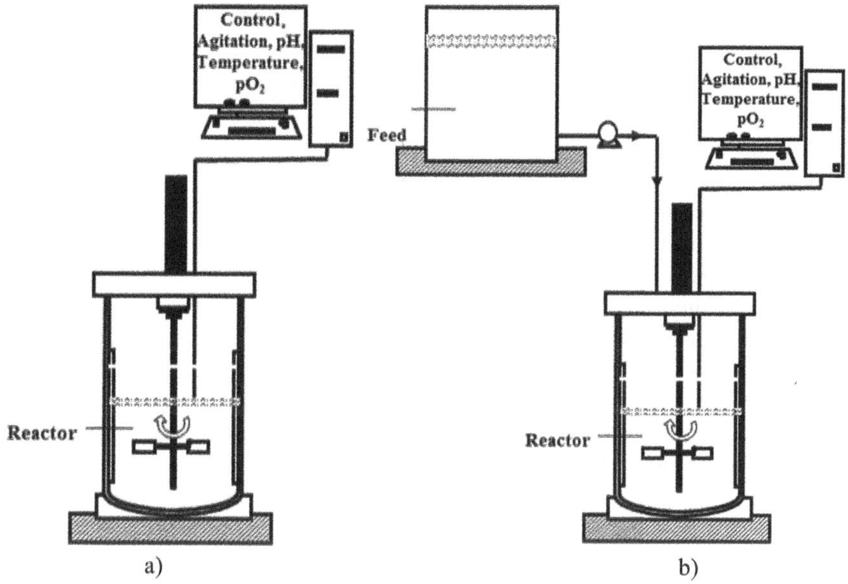

Figure 1.7.a. *Conventional bioreactor culture processes: a) batch culture; b) fed-batch culture*

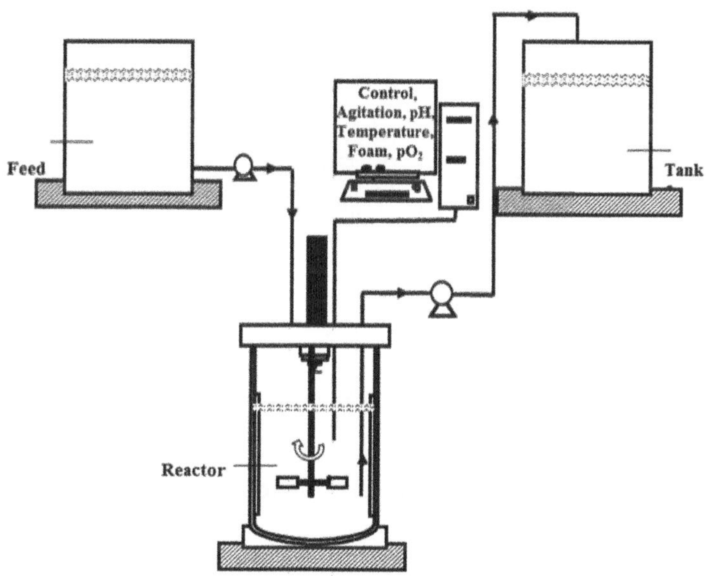

Figure 1.7.b. *Conventional bioreactor culture processes: continuous culture (continued)*

In order to ensure process scaling up (Figure 1.8), a number of equations making use of adimensional figures and the theory of similarities have been developed. The utilization of these equations requires that a standard reactor geometry and their agitation system be satisfied (see Chapter 5 of the first volume). The main adimensional numbers considered are Reynolds (action of viscosity forces), power (action of inertia forces), Froude's (gravity force action), and Weber's numbers (action of surface tension forces), etc. However, scaling up does not make it possible to satisfy the constraints of the process in an equivalent way according to the volume. This leads to a choice for engineers that will have consequences on the performance of their large-scale processes. A good example is the need to maintain a high oxygen transfer while limiting shear forces harmful to microorganisms. This dilemma is particularly acute for filamentous microorganisms that are most often more sensitive to shear forces. These difficulties associated with upscaling have, in a number of cases, resulted in abandoning very promising small-scale projects that are unfortunately difficult to transfer to larger scales.

Figure 1.8. *Upscaling of the bioreactor culturing process (200 ml to 2 m^3)*

1.5.3. *Downstream fermentation processes (downstream processing, DSP)*

Downstream fermentation processes are dependent on the nature of the product to be recovered (cells, intracellular or extracellular metabolites) and its application.

A first solid-liquid separation step is always required. As a general rule, it makes use of sedimentation, accelerated (centrifugation) or not, or filtration. When the product to be recovered is intracellular, an operation for bursting cells is necessary: their destruction may be mechanical, by solid (namely grinding) or liquid (such as pressure difference), or non-mechanical constraints (physical, chemical or enzymatic lysis).

Subsequently, a number of unit operations also used in chemical engineering (see Chapter 5 of the first volume) can be chained up to yield a more or less pure product in dry or non-dry form (extraction, precipitation, ultrafiltration, chromatography, drying, etc.).

1.5.4. *Coupled procedures*

The first unit purification operations can, in a number of processes, be directly coupled to production and thereby improve productivity. This may include, for example, the continuous extraction of a compound that is toxic to the cell or likely to be rapidly degraded in the culture medium. A fine example of a coupled process is the membrane-based bioreactor (Figure 1.9). In such a device, the reaction medium is in permanent contact with a filter membrane (immersed into the bioreactor or connected through an external circulation loop) that allows the continuous extraction of the culture medium free of its cells (microfiltration) or of the latter as well as molecules larger than the cut-off threshold of the membrane (ultrafiltration).

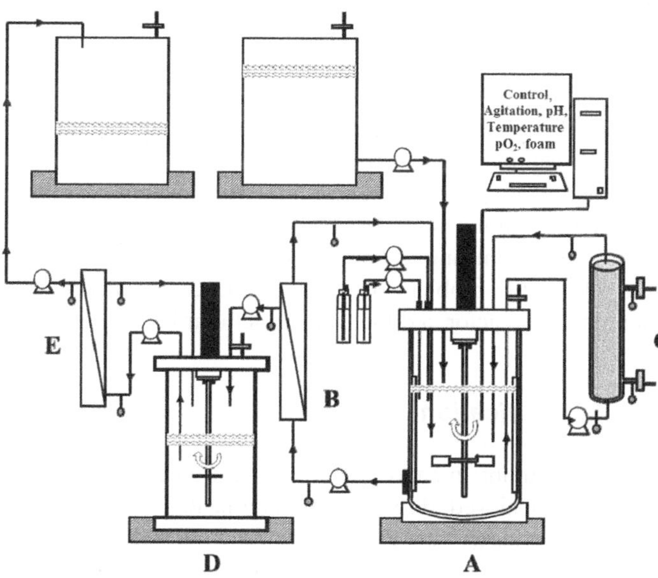

Figure 1.9. *Coupled membrane process for biosurfactant production*

COMMENT ON FIGURE 1.9.– *Microorganisms are cultured in a first reactor (A) connected to a microfiltration membrane (B). The latter allows the extraction of the biosurfactant produced continuously. The bioreactor aeration is achieved by way of an air-liquid membrane contactor (C) which severely restricts the presence of foam due to the biosurfactant (absence of bubbles in the bioreactor). Finally, the microfiltrate recovered from an ancillary tank (D) is purified by ultrafiltration on a membrane (E) with a cut-off threshold of 10,000 daltons that lets small molecules pass through, but not micelles formed by the biosurfactant.*

1.5.5. *Microfluidic* intake (scale-up/scale-down)*

Microfluidics is gradually invading the industrial microbiology sector. Its use has several vocations: to miniaturize bioreactors to facilitate high-throughput studies of optimal culture conditions or the screening of mutants as mentioned above, to study the behavior of cells individually and to collect the most efficient cells.

The development of **optodes**, which are miniaturized sensors based on the emission of a light source according to the variation of a culture parameter such as pH or dissolved oxygen concentration, has enabled the development of miniature bioreactors of a few hundred microliters. In addition, advances in molecular biology are now generating genetically modified strains in which genes encoding fluorescent proteins are placed downstream of a promoter involved in the expression of a gene of interest. This "reporter gene" logic is a means to follow the expression of a given gene in different circumstances by a global measure of the fluorescence emitted by all cells in culture. The behavior of each cell can be analyzed using flow cytometry. This technique, also used to count cells in culture, enables the analysis of the morphology or of the phenotypic behavior of every cell by making them pass through a very fine capillary in the beam of a laser. Microfluidics, in connection with time-lapse microscopy, allows one to follow the development of a colony from a single cell and the evolution of a possible phenotypic heterogeneity based on this mother cell. These very promising technologies have made it possible, in recent years in particular, to add to the problem of physical and chemical heterogeneity of the cell environment in a bioreactor, a dimension of cellular heterogeneity.

1.5.6. *Process intensification*

The scaling-up constraints mentioned above, the progress of small-scale tools and the high productivity achieved through cell optimization are likely to completely revolutionize the bio-industry of the future. One of the solutions increasingly envisaged is the multiplication of small volume production tools, operated in parallel and in which process optimization has been validated. This approach could increase the success rate of transferring bioprocesses from the pilot lab scale to the company's scale.

1.6. Innovative concepts

1.6.1. *Biofilm reactors*

The development of immobilized cell reactors is a fairly old concept that has been used for many years in industrial microbiology. In a number of cases, these reactors provide increased productivity for a metabolite or enzyme of interest. They also facilitate the development of continuous processes for the permanent extraction of a culture medium virtually devoid of cells. Conventional approaches used are fixation on a support by adsorption, encapsulation in gel or the exploitation of the natural or induced formation of cellular aggregates. In recent years, the discovery and characterization of biofilm growth of a number of microorganisms has led to the development of a new approach to processes that incorporate this differentiated form of growth. Although still complex to control, the formation of biofilm provides a natural form of cellular immobilization that responds in nature to a number of stresses. In the long run, biofilms may eventually be of interest, particularly in the production of certain secondary metabolites.

1.6.2. *Mixed cultures and cascades of microorganisms*

For many years, the logic of pure, perfectly controlled culture has predominated in most industrial microbiology processes. Advances achieved in microbial ecology and in characterizing microbial ecosystems* are paving the way for a new era that could eventually create more resilient processes, integrating a microbial community rather than a pure microorganism. Another evolution in microbial processes is the development of cascades of microorganisms with, in particular, different but complementary enzymatic capabilities for obtaining a particular product.

1.7. Towards a digital bio-industry

The biodiversity of microorganisms and the complexity of their operating mode are driving forces underlying the development of digital technology in bio-industry (see also Chapter 5 of this volume on the impact of digital transition on industry). Both involve the creation of databases that gather information collected due in particular to the explosion of genomics*, proteomics*, transcriptomics*, etc. These data also constitute a source that feeds into numerous increasingly complex models based on biostatistical analyses or learning techniques. It should lead to the possibility of predicting the behavior of a microorganism in a given environment and thus controlling that behavior. This is notably the subject of **systemic biology*** which attempts to include as much information available as possible into these models (Figure 1.10).

Degree of importance of digital technology →

Screening		
Screening on solid and liquid media	High-throughput screening	Screening *in silico*
Strain optimization		
Genetic engineering	Metabolic engineering	Synthetic biology
Optimization of culture conditions		
Screening in vials and experimental planning	Microreactors	Systems biology
Process optimization		
Upscaling	Cell scale monitoring	Parallelization and microfluidics

Figure 1.10. *Evolution of approaches to industrial microbiology and the significance of digital technology*

1.8. Acknowledgements

The author wishes to thank the REALCAT platform of the École centrale de Lille and the University of Lille for the photos of the robots, the bio-driver and the colony picker. The platform received a national grant managed by the *Agence nationale française de la recherche* (ANR – French National Research Agency) as part of the *Programme d'investissements d'avenir* (PIA – Program of investments for the future), under the contractual reference ANR-11-EQPX-0037. The Hauts-de-France region, the FEDER, the École centrale de Lille and the Fondation centrale initiatives are greatly thanked for their financial contribution to the acquisition of the equipment of the REALCAT platform.

1.9. Glossary

Archaebacteria

Or *Archaea*, formerly known as archaeobacteria, are prokaryote single-celled microorganisms, or more specifically living beings composed of a single cell that does not include a nucleus or organelles, like bacteria.

Auxotrophy

The inability of a living organism to synthesize an organic compound necessary for its development. that is characterized by this defect. Conversely, the term prototrophy is employed.

Bacteria

Single-celled organisms, with a very simple structure, devoid of nucleus and organelles, with diffuse genetic material, usually without chlorophyll and reproduces by scissiparity.

Biopesticide

A living organism or one of its derivatives capable of controlling the activity of a pathogen organism or that is harmful to plants. It works by reducing inoculum and/or interfering with one or more stages of the infectious cycle of the phytopathogen.

Bioprocess

Bioprocesses refer to, according to the most widely accepted and shared definition in the field of biotechnology, all the implementations of living systems, or their constituents, for the production of knowledge, goods or services.

Bioremediation

A technique consisting of increasing biodegradation or biotransformation, by inoculating specific microorganisms (bio-augmentation) or by stimulating the activity of indigenous microbial populations, by biostimulation, by nutrient intake and by adjusting the conditions of the medium.

Biostimulant

In the areas of human or animal health and agriculture, a biostimulant or plant defense stimulant (PDS) is a product for the preventive stimulation of the activity of an organism (vegetable in the case of PDS) or for the stimulation of its immune system.

Biosurfactant

An active biological compound of detergents that dissolves or suspends stains adhering to surfaces. It promotes humidification, solubilization and the emulsion of organic compounds.

Chemolithotroph

A chemolithotrophic organism, typically a bacterium, uses accordingly an organic molecule or mineral compound as energy source (chemosynthesis) and oxidizing mineral compounds such as NO_3 nitrates, O_2 dioxygen, etc., as electron donors.

Chemoorganotroph

An organism that derives its energy from the metabolism of an organic substrate (sugar, amino acid, fatty acid, etc.).

Deoxyribonucleic acid (DNA)

A nucleic acid characteristic of chromosomes, consisting of two strands coiled in double helix and each formed by a succession of nucleotides. As a carrier of genetic information, DNA ensures the control of cell activity.

Ecosystem

A system formed by an environment (biotope) and by the set of species (biocenosis) that live, feed and reproduce therein.

Enzyme

A protein that accelerates chemical reactions.

Fungi

A group that includes single-celled eukaryotic organisms (yeasts) or multicellular organisms (molds) of microscopic size as well as macroscopic fungi.

Gene

A segment of DNA that conditions the synthesis of one or more proteins and thus the manifestation and transmission of a specific hereditary character.

Gene cluster

A set of genes encoding enzymes involved in the synthesis of the same molecule.

Gene expression

Refers to all of the biochemical processes by which the hereditary information stored in a gene is read to lead to the production of molecules that will have an active role in cellular functioning, such as proteins.

Genomics

Genomics is the science of genomes. It studies DNA sequences in living beings. A genome is made up of the set of genetic information contained in the cell. For example, in a human cell, the genome is composed of information carried by the 23 pairs of chromosomes of the nucleus as well as by the DNA present in mitochondria. Genomics helps, for example, to better understand the diversity of life, build phylogenetic trees or identify genes associated with diseases.

In silico

Standing for "in silicon", *in silico* is a Latin-inspired neologism designating research or trials performed by means of complex computerized calculations or computer-based models. This expression is mainly used in the fields of genomics and bioinformatics.

In vitro

Used to describe an experiment carried out in an artificial medium, in the laboratory.

Lactic acid ferments

Lactic acid ferments are bacteria used for processing milk into dairy products, such as yogurt and cheese, but also for processing other foods.

Metabolite

A metabolite is an intermediate organic compound or it originates from the metabolism. This term is generally reserved for small molecules and monomers, as opposed to macromolecules. Therefore, unlike glycogen, glucose is a metabolite which is a polysaccharide of very high molecular weight.

Metabolomics

Metabolomics is a very recent science that studies all primary and secondary metabolites in the case of plants present in a cell, organ or organism. This is the equivalent of genomics for DNA.

Microalgae

The term microalgae, sometimes called a microphyte, refers to microscopic algae. Microalgae have been consumed around the world for thousands of years. For example, traces of the consumption of various species of microalgae in Mexico during the Aztec period have been found.

Microbial chassis

A model microorganism genetically manipulated to be able to produce different types of heterogeneous metabolites.

Microfluidics

Microfluidics is the science and technology of fluid-handling systems, of which at least one of the characteristic dimensions is in the order of a micrometer.

Miniaturization

The creation of mechanical, optical or electronic products and their devices at increasingly smaller scales.

Mold

A small-sized fungus that causes a chemical modification in the medium on which it grows.

Mutagenesis

Action of modifying the sequence of a gene.

Petri dish

A box of glass or plastic material used in bacteriology for culturing on solid media.

Phenotype

In genetics, the phenotype is the set of all observable characteristics of an organism. Very often, the use of this term is more restrictive: the phenotype is considered at the single character level, at the cellular or even molecular level.

Photo-organotrophic

Photo-organotrophic organisms are photosynthetic bacteria that utilize organic molecules as a carbon source, an organic molecule as an electron source for reducing this carbon, and light as an energy source (some cyanobacteria, for example).

Promoter

A genetic sequence located downstream of a gene and that manages its expression.

Proteomics

The study of proteins existing in a cell or tissue, for the purpose of identifying those that are, for example, specific to a pathology. Its applications concern in particular medical diagnostics, biotechnological and pharmaceutical research.

Respiration

A set of phenomena that enable oxygen absorption and the release of carbon dioxide from living beings.

Robotization

The substitution of human operators by robots for performing industrial tasks. The action of replacing someone's reasoned behaviors with automatisms.

Screening

Techniques in the field of pharmacology, biochemistry, genomics and proteomics techniques that aim to study and identify in chemical libraries and target libraries, molecules biologically active that present new properties.

Strains

A set of cells derived from the same clone and thus presenting the typical characteristics of the original bacterium or virus.

Synthetic biology

Synthesis biology, or synthetic biology, is an emerging scientific and biotechnology field that combines biology and engineering principles, with the aim of designing and building new living organisms.

Synthon

A "fictitious molecule" whose representation symbolizes reactivity. This concept is used when planning for the retrosynthesis of complex organic compounds. They are derived from a simple fragmental decomposition of the molecule which is to be assembled. In biotechnology, synthons are "building blocks" that could be used for the biosynthesis of biopolymers.

Systemic biology

The biology of systems (or systemic biology, or integrative biology) is a recent field of biology that studies living organisms as systems since that is what they actually are, as opposed to historical approaches that tend to break down the study at every level, into biology, physiology, biochemistry, etc.

Transcriptomics

A study of all messenger RNAs produced during the process of genome transcription. It relies on the systematic quantification of these mRNA, which makes it possible to obtain a relative indication of the transcription rate of different genes under given conditions.

Yeast

Eukaryotic microorganism member of the fungus kingdom. The most famous industrial yeast, *Saccharomyces cerevisiae*, is mainly used for its ability to transform carbohydrates into carbon dioxide and alcohols. It is used for baking and the production of alcoholic beverages.

1.10. References

Baltz, R.H., Devies, J.E., Demain, A.L. (2010). *Manual of Industrial Microbiology and Biotechnology*. ASM Press, Washington.

BBC Research (2018a). World market for fermentation ingredients. Report, February.

BBC Research (2018b). Food Enzymes. Global market. Report, August.

BBC Research (2018c). Global market for enzymes in industrial applications. Report, September.

Hill, D., Sugrue, I., Arendt, E., Hill, C., Stanton, C. and Ross, R.P. (2017). Recent advances in microbial fermentation for dairy and health. *F1000Research* [Online]. Available at: https://f1000research.com/articles/6-751.

2

Hydrogen Production by Steam Reforming

2.1. The industrial production of hydrogen

2.1.1. *The processes of hydrogen production*

2.1.1.1. *History*

In 1868, two French scientists – Tessié du Motay and Maréchal – were the first to describe a process for converting hydrocarbons into hydrogen in the presence of water. However, hydrocarbon steam reforming did not generate industrial interest in hydrogen production until 1924, when Neumann and Jacob published the first detailed study of the catalytic reaction between methane and steam. Many patents were registered in the years that followed and the first industrial steam reformer was used in 1931 in Baton Rouge, Louisiana (USA) by the Standard Oil Company of New Jersey, now Exxon. Further implementations followed rapidly, mainly in England and Germany. It was not until 1946 that the first French reformer was used in Toulouse by the ONIA (*Office National Industriel de l'Azote*). Over the following decade, developments in metallurgy enabled the implementation of pressurized steam reforming, thus improving the energy efficiency of the process through better heat recovery and lower compression costs in ammonia and methanol production units. In the 1960s, the reforming of heavier hydrocarbons, such as naphtha, was developed (Rostrup-Nielsen 1983). Today, hydrogen production by steam reforming is a mature process. Hydrogen is mainly produced by natural gas steam reforming, since the latter is widely available, easier to reform and allows better energy efficiency.

Chapter written by Marie BASIN, Diana TUDORACHE, Matthieu FLIN, Raphaël FAURE and Philippe ARPENTINIER.

There are other existing technologies based on the partial oxidation of heavier feeds (oil residues, deep-cuts, etc.): autothermal reforming combining a steam conversion with oxygen conversion or partial oxidation. These technologies require the production of pure oxygen using an air gas separation unit; their investment is thus more expensive. After launching in the 1980s, these technologies remain marginal, especially for hydrogen production.

The purpose of this chapter is to study the various production and processing operations of synthesis gas and hydrogen using reforming, though some units for this kind of processing also apply to facilities producing synthesis gas by partial heavy fuel oxidation or coal gasification. This is the case for conversion and decarbonization units. Some of these technologies are presented and compared with the steam reforming process in the next section.

2.1.1.2. *Comparison of processes*

As explained above, there are three processes for hydrogen production:

– steam reforming (SR) consists of the catalytic conversion of hydrocarbons using steam as an oxidizing agent;

– autothermal reforming (ATR) is a combination of steam reforming of hydrocarbons on a catalytic bed with partial oxygen oxidation;

– partial oxidation (POX) is based solely on non-catalytic conversion of hydrocarbons with oxygen.

These processes generate a gas called synthesis gas: this is a mixture of methane, hydrogen, carbon monoxide, carbon dioxide and nitrogen. Depending on the reactions involved and the operating temperatures of these reactors, the composition of the synthesis gas produced varies, in particular the H_2/CO ratio.

Steam reforming, which will be further detailed hereafter, converts hydrocarbons into synthesis gas using steam. Since these reactions are highly endothermic, they are carried out in a furnace providing the heat needed for reaction to catalyst-filled tubes.

Autothermal reforming combines a partial oxidation section in a burner and a steam conversion section on a catalytic bed; the first section providing the necessary heat to the second, hence its name "autothermal". This technology can be used on its own or in combination with a steam reformer. In the case of hydrogen production, this process requires the investment of an air-gas separation unit, in order to provide pure oxygen at the reactor inlet.

On the other hand, partial oxidation consists of oxygen conversion of light to heavy hydrocarbon feeds, ranging from natural gas to petroleum coke, in a reactor equipped with refractory lining. As with the autothermal reformer, this reactor requires the investment of an air-gas separation unit. Steam can also be injected into the reactor to control its temperature, but there is no specification regarding a minimal level of inlet steam.

The characteristics of these reactors are summarized in Table 2.1. An examination of this table helps in understanding why the steam reforming process is the preferred process for hydrogen production: it presents the highest H_2/CO ratio; its range of production capacity is wide and does not require investing in a complementary unit for oxygen production.

	SMR	**ATR**	**POX**
Catalytic process	Yes	Yes	No
Need for oxygen	No	Yes	Yes
Feedstock	Light hydrocarbons	Light hydrocarbons	Light and heavy hydrocarbons
Maximum pressure (bar)	45	100	80
Maximum outlet temperature (C)	950	1,050	1,400
H_2/CO outlet ratio of the reactor	~3–5	1.6–2.2	1–2
Product	H_2 (CO)	H_2 (CO)	CO (H_2)
Unit size (Nm^3/h)	10,000–200,000	Up to 1,000,000	Up to 150,000

Table 2.1. *Comparison of the characteristics of hydrogen production processes*

2.1.2. *Natural gas steam reforming*

2.1.2.1. *Process overview*

The steam reforming process converts hydrocarbons into a mixture of hydrogen, carbon monoxide, carbon dioxide and methane through highly endothermic reactions known as Steam Reforming. The reaction of carbon monoxide with steam, better known as the water gas shift reaction, also takes place in parallel and converts carbon monoxide into carbon dioxide and hydrogen, in the presence of steam. Hydrocarbon steam reforming requires a substantial amount of heat; this is why the reaction is carried out in a furnace (Figure 2.1) where the combustion of natural gas and tail gas from the unit generates the necessary heat.

Figure 2.1. *Hydrogen production unit using steam reforming*

The process comprises the following steps (Figure 2.2):

1) natural gas pre-treatment by hydrodesulfurization to remove sulfur compounds from the feedstock and avoid catalyst poisoning;

2) reforming of heavy hydrocarbons in a pre-reformer to prevent these heavy hydrocarbons from cracking inside the main reactor (optional);

3) reforming in a furnace that can hold several hundred tubes and burners, the tubes being about three times more in number than the burners;

4) carbon monoxide conversion into hydrogen and carbon dioxide by the Water Gas Shift reaction;

5) purification of the gas mixture to produce high purity hydrogen (molar fraction > 99.99 %) in a pressure inversion adsorption/desorption unit, also known as a PSA (Pressure Swing Adsorption) unit; tail gas from this unit is used as fuel.

These different steps will be presented in more detail in the following sections.

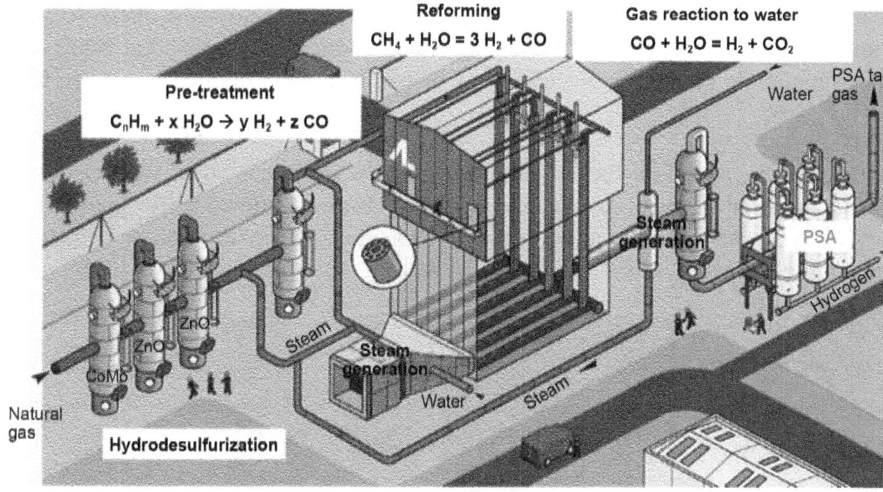

Figure 2.2. *Block diagram of a steam reforming hydrogen production unit. For a color version of this figure, see www.iste.co.uk/dalpont/process2.zip*

2.1.2.2. *The pre-treatment of natural gas*

2.1.2.2.1. Hydrodesulfurization

Natural gas contains sulfur compounds that can reach levels of one or more tens of ppm vol. These sulfur compounds are usually hydrogen sulfide (H_2S), light mercaptans (R-SH) or odorous tracers such as tetrahydrothiophene (THT) for natural gas delivered from the distribution network. These compounds are poisons for the catalysts implemented in reactors located downstream. The desulfurization of natural gas allows the near-total elimination of sulfur present to reach the maximum allowable residual content of 0.1 ppm vol (Giroudière and Le Gall 2016). Hydrodesulfurization is achieved in two stages:

– hydrogenation of sulfur compounds: their conversion into H_2S by catalytic reaction in the presence of hydrogen;

– hydrogen sulfide trapping on an adsorbent made of zinc oxide.

The hydrogenation stage takes place in an adiabatic reactor, shown in Figure 2.3, operated at a temperature between 350°C and 400°C in the presence of a catalyst containing molybdenum and either cobalt (Co-Mo) or nickel (Ni-Mo). This reaction occurs in the presence of hydrogen usually recycled from the production. The hydrogen content at the reactor inlet is between 2% and 5% in volume. The choice of catalyst depends on the feedstock to be treated: Co-Mo catalysts have a superior activity but Ni-Mo catalysts are recommended for feeds containing carbon monoxide because of their low sensitivity to it. The lifetime of these catalysts is about 10 years (Giroudière and Le Gall 2016). The implementation of hydrogenation catalysts is rather simple: the catalyst is placed in one or two beds in cylindrical reactors (fixed beds).

The hydrogen sulfide fixing stage is generally implemented in two adsorption reactors installed in series, the second operating as an additional sulfur guard in the event of hydrogen sulfide leaking in the first adsorber. These reactors are operated at temperatures between 350°C and 400°C as well.

Figure 2.3. *Hydrodesulfurization unit consisting of a hydrogenation reactor (right-hand side) and two adsorption reactors in series (on the left)*

2.1.2.2.2. Pre-reforming of natural gas

Desulfurized natural gas can be sent either directly into the steam reformer or into an optional reactor called a "pre-reformer". The objective of this reactor is to convert the heaviest hydrocarbons of natural gas (up to ethane) into a mixture of methane, hydrogen, CO and CO_2. This reactor has several advantages:

– it acts as protection for the steam reformer in the event of a malfunction of the hydrodesulfurization unit;

– it also protects the steam reformer by avoiding the cracking of heavy hydrocarbons and therefore allows the processing of more or less heavy variable feeds;

– the size of the reforming furnace can be reduced, and so too its price: the pre-reforming reactor, consisting of a vertical cylindrical container with a fixed bed of catalyst, is relatively inexpensive;

– variations in the composition of the feed (natural gas) can be smoothed, especially if there are several sources;

– the total steam production of the process can be reduced (see section 2.1.2.3.5 on thermal integration of the unit).

The pre-reformer is a fixed-bed catalytic adiabatic reactor, illustrated in Figure 2.4. The hydrocarbon conversion reaction takes place in the presence of a nickel-based catalyst. The reactor inlet temperature is between 450°C and 480°C. In the case of natural gas, the reactions involved are endothermic, therefore the outlet temperature is between 380°C and 430°C.

Figure 2.4. *Example of a pre-reformer*

2.1.2.3. *Steam reforming technologies*

2.1.2.3.1. Thermodynamics and kinetics of the reaction

As presented in the introduction to this chapter, the steam reforming process converts hydrocarbons into a mixture of hydrogen, carbon monoxide, carbon dioxide and methane, by reactions whose generalized equation is as follows:

$$C_nH_m + n\ H_2O = n\ CO + (n + m/2)\ H_2 \qquad [2.1]$$

Methane is the main component of natural gas and if a pre-reformer is present, it is the only hydrocarbon entering the steam reformer. The steam reforming reaction of methane is therefore the main reaction occurring:

$$CH_4 + H_2O = CO + 3H_2,\ \Delta H_{760°C} = 226\ kJ/kmol \qquad [2.2]$$

Reaction [2.2] is endothermic and thermodynamically balanced. Methane conversion, and therefore forward reaction, will be enhanced at high temperatures. In addition, the number of moles of products is greater than the number of reactant moles. Hence, operation at low pressure favors methane conversion (Le Châtelier principle).

The water gas shift reaction also takes place in parallel and converts carbon monoxide into carbon dioxide and hydrogen in the presence of water:

$$CO + H_2O = CO_2 + H_2, \Delta H_{760°C} = -34.3 \text{ kJ/kmol} \quad [2.3]$$

Reaction [2.3] is exothermic, thermodynamically balanced and equimolar. CO conversion and forward reaction are therefore favored at low temperatures and pressure has no influence on thermodynamic equilibrium (Giroudière and Le Gall 2016).

Finally, reactions [2.2] and [2.3] are catalyzed by nickel catalysts (see section 2.1.2.3.3) to increase their reaction rate. These catalysts are in solid form, insoluble in the reaction medium, and the catalytic reaction occurs on the surface of the solid.

It is therefore a heterogeneous catalysis that can be described in the form of seven key steps (see Figure 2.5):

– the diffusion of the reactant(s) through the gaseous layer stagnating around a grain (external diffusion, Step 1);

– the diffusion of the reactant(s) into the grain pores (internal diffusion, Step 2);

– chemisorption of reactants at active sites (Step 3);

– the conversion of the reactants into products (superficial reaction, Step 4);

– the desorption of products from active sites (Step 5);

– the diffusion of the products into the grain pores (internal diffusion, Step 6);

– the diffusion of the products through the gaseous layer stagnating around a grain (external diffusion, Step 7).

The stages for mass transfer may be slower than the reaction itself: they then limit the reaction rate. As early as 1982, studies have shown the impact of diffusional limitation of reactants on the reaction rate. The calculated efficiency factors are in the range of 0.05 to 0.08 (De Deken *et al.* 1982; Soliman *et al.* 1988), meaning that at least 95% of the catalyst fed into the reactor is not used to catalyze the reaction because reactant molecules do not have time to reach active sites.

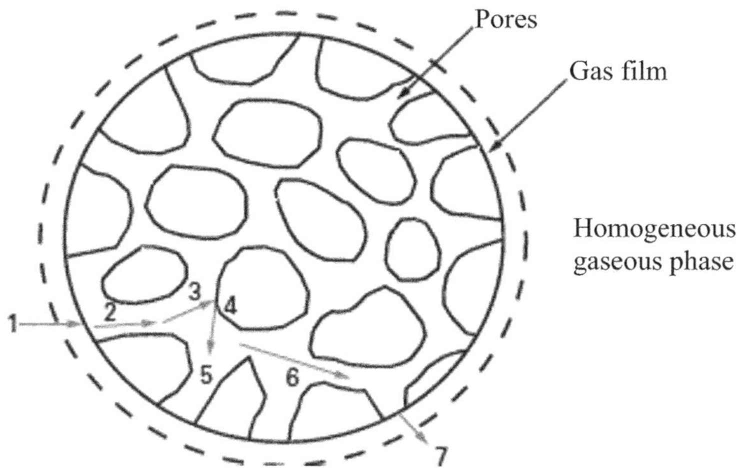

Figure 2.5. *The seven stages of heterogeneous catalysis (Guisnet and Naccache 2004)*

2.1.2.3.2. The design of steam reforming furnaces

Since the main reforming reaction is endothermic, it requires an external heat input that is achieved industrially by combustion of natural gas and tail gas from the PSA. The steam reforming furnace is thus composed of tubes containing the catalyst, essential for the reaction, and in which the reactants flow, arranged vertically in a combustion chamber. In industrial combustion chambers, the heating system consists of burners that release about 2 MW of power each.

The conversion in the steam reformer is driven by the thermodynamic equilibrium of the methane steam reforming reaction and therefore depends on:

– the reactor outlet temperature: the higher the outlet temperature, the better the conversion;

– the operating pressure: the lower the pressure, the better the conversion;

– the ratio of the steam molar flow rate to the molar flow rate of the carbon contained in the feed (calculated by only taking into account hydrocarbons and not the CO or CO_2 content), usually called S/C or steam-to-carbon ratio: the higher the ratio, that is, the more steam present in excess, the better the conversion.

The choice of operating conditions leading to the overall performance of the procedure is a trade-off, achieved by considering the following points:

– the higher the operating temperature, the more specific the metallurgy of the tubes must be and this can be costly: the reforming temperature is limited to about 920°C;

– 1) natural gas is available from the high-pressure network, 2) synthesis gas purification requires a minimum inlet pressure between 10 and 15 bar and 3) the hydrogen produced is generally used under pressure: although pressure close to atmospheric pressure allows for near-total conversion, it is more beneficial to minimize compression costs by implementing the process at a pressure between 15 and 45 bar;

– the S/C ratio must be kept above a threshold value (usually 1.7) to avoid carbon formation (see section 2.2) and the higher the S/C ratio, the higher the H_2/CO ratio in the synthesis gas.

The operating conditions induce a temperature profile within the reforming tubes. There is a significant gradient between the top of the tube, which is typically at temperatures between 650°C and 700°C (corresponding to the inlet temperature of the feed gas mixture) – and the bottom of the tube – subjected to temperatures between 900°C and 950°C (corresponding to the outlet temperature of the synthesis gas).

In addition, the reforming tubes are designed to withstand a significant pressure gradient: the process gas usually has a pressure between 15 and 45 bar while the fumes circulate at a pressure slightly below the atmospheric pressure. Indeed, the burners are operated at low pressure, a few mbar below the atmospheric pressure.

Existing reformers that can achieve such temperature levels have been designed with a wide variety of tubes and burner arrangements.

As shown in Figure 2.6, burners can be implemented:

– at the top level of the furnace, a configuration better known as top-fired. These vertical furnaces can consist of one or more rows of tubes with rows of burners arranged between rows of tubes, as well as between rows of tubes and refractory walls. This technology was originally developed by ICI (Imperial Chemical Industries) and many other companies have sold furnaces of this type (KTI, Uhde, Technip, Air Liquide-Lurgi, etc.);

– at the floor level, a configuration better known as bottom-fired. CdF Chimie (now Exxon) originally industrialized it – is the least widespread now;

– laterally, on the sides of the furnace themselves, better known as side-fired. This configuration was developed by Haldor Topsoe. These furnaces consist of a row of tubes with small radiant burners on two walls, distributed over several heights;

– laterally, on the sides of the furnace but in small terraces, better known as terrace wall. Foster Wheeler furnaces are based on this arrangement comprising two or three levels of burners.

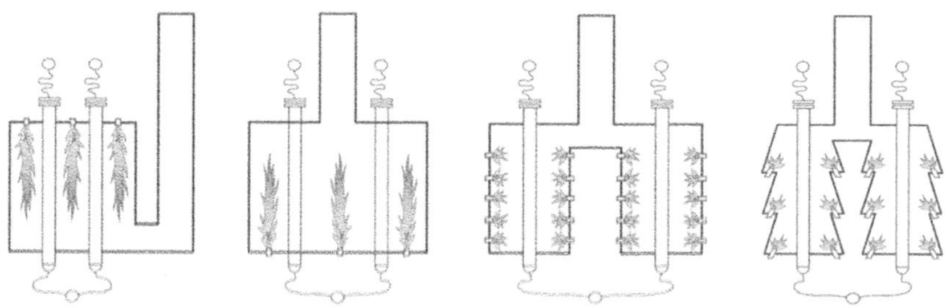

Figure 2.6. *Examples of the design of steam reforming furnaces (Giroudière and Le Gall 2016)*

The design of the tubes (section, length, thickness) influences the transfer of heat to the catalyst bed. The length of the tube is sized to obtain the necessary exchange surface: the determining designing phenomenon is therefore the thermal transfer and not the reaction kinetics. Tube dimensions are typically as follows: the internal diameter ranges between 100 mm and 125 mm, the thickness of the wall lies between 10 mm and 15 mm and the height is between 10 m and 15 m. The tubes used are still centrifuged tubes made of high alloy steel (nickel base). The materials of the tubes should have the best possible resistance to high-temperature creep.

The tube must therefore first withstand high pressure and temperatures, but also act as a heat transmission medium. Inside, the tube contains a mixture of gaseous compounds (CH_4, CO, CO_2, H_2O and H_2) and outside it is exposed to oxidizing combustion products. The thickness of the tubes cannot be increased, it could reduce heat transfer and increase thermal stresses. The joints are incompatible with very severe cycles of temperature and mechanical effort; the materials used must therefore be weldable so as to ensure perfect sealing under high temperatures and high pressure.

Videos
Virtual tour of a top-fired steam reforming furnace: https://youtu.be/jlishHLlGzE

Steam reforming furnaces consist of two distinct areas:

– a "radiation" zone, where tubes and burners are placed. The predominant mode of heat transfer in this area is the radiation of flames, hot fumes and refractory walls;

– a "convection" zone: at the outlet of the previous area, the fumes are at temperatures around 1,000°C and exchangers are integrated to allow the recovery of the heat contained in these fumes and to enable preheating of various feeds (natural gas, steam, combustion air). The predominant heat transfer mode here is convection.

A typical arrangement is given in Figure 2.7. Depending on the operating conditions and the design of the furnace, 45% to 55% of the heat released by combustion at the burner level is transmitted to the reforming tubes in the radiation area. The remaining heat is recovered in the convection zone.

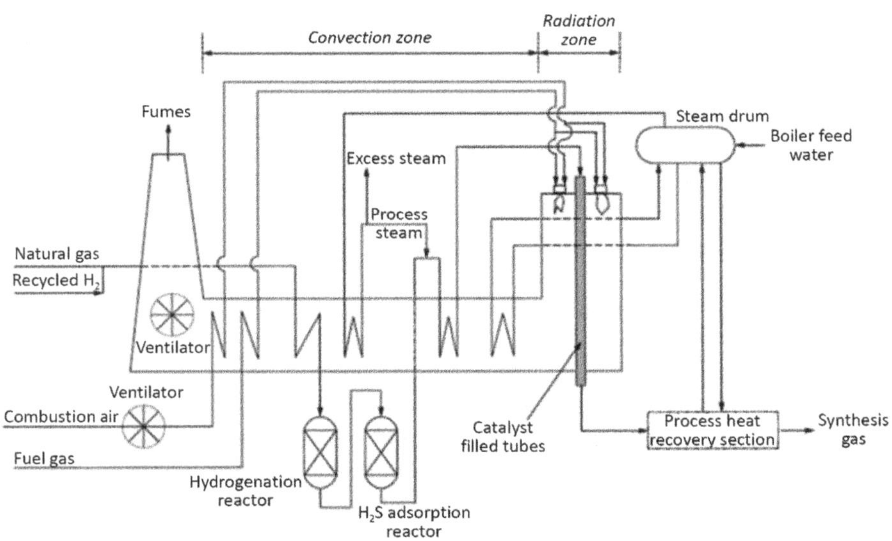

Figure 2.7. *Arrangement of convection and radiation areas inside a steam reforming unit (Giroudière and Le Gall 2016)*

2.1.2.3.3. Formulation and production of steam reforming catalysts

Since the kinetics of the steam reforming reaction are limited, catalysts are used in order to achieve thermodynamic equilibrium more quickly. Only catalysts used industrially are described here.

Catalyst formulations

The most commonly used catalysts are composed of Ni at 10% to 15% w/w, dispersed over a refractory medium with a calcium-doped alumina base ($CaAl_4O_7$, $CaAl_{12}O_{19}$) or a magnesium alumina base ($MgAl_2O_4$). Figure 2.8 illustrates a Clariant Reformax 330 catalyst designed for natural gas steam reforming.

Figure 2.8. *A typical catalyst for natural gas steam reforming (Clariant Reformax 330)*

Hydrocarbon molecules are adsorbed and dissociated at metal sites, while water is adsorbed at metal oxide sites. Adsorbed water can then migrate to active metal sites to react with dissociated hydrocarbons. Nickel is not the most active metal for steam the reforming reaction but it has a satisfactory activity while maintaining a relatively low cost. Nickel is on average 3,000 times cheaper than rhodium, which is considered the most active metal for the steam reforming reaction (Table 2.2).

	Rh	Ru	Ni[b]	Ir	Pd	Pt
Comparative activities[a]	1.6	1.4	1	0.7	0.6	0.5

a: comparative activities for silica-supported catalysts (SiO_2). Operating conditions: molar ratio $H_2O/CH_4 = 3$, 500°C, 0.1 MPa.

b: Ni is the reference catalyst, in other words, activities are standardized with respect to that of Ni.

Table 2.2. *Comparative activities of a few metals in methane steam reforming reaction*

Nevertheless, the order of activity shown in Table 2.2 may be moderate. As a matter of fact, the activity of metals is subject to changes according to experimental parameters, such as temperature and pressure. Regarding the supported catalysts used for toluene steam reforming, it was shown that the activity also depends on the catalyst support. The ceramic support of the catalyst plays a key role. It has to be chemically and thermally inert in order to avoid disabling the catalyst by sintering. In addition, the support must also present adequate acid-base properties. Therefore, as part of the steam reforming reaction, an acidic support stimulates carbon formation. Several supports can be classified in order of decreasing acidity:

SiO_2-Al_2O_3 > SiO_2-MgO > SiO_2 > **α-Al_2O_3** > TiO_2 > ZrO_2 > **$MgAl_2O_4$** > UO_2 > **CaO, MgO**

It appears that MgO-, CaO- or MgA12O4-based ceramic supports are less acidic than α-Al_2O_3 alumina. In addition, MgO-, CaO- and A12O3-based supports are thermally very stable, including under high water steam partial pressure. These supports are therefore well adapted to this process. However, additives are frequently added to the formulation of the catalyst support or during the impregnation of active phases. Alkaline metals, such as potassium, are mainly used, which alters the acid-basicity of the support and prevents coke formation.

Production of methane steam reforming catalysts

These catalysts are produced by extrusion or pelletization of metal oxide mixtures and then by impregnation with solutions of precursors of active metals. A mixture of precursor powders of the desired material such as boehmite ($Al(OH)_3$) and magnesium carbonates ($MgCO_3$) which also include, according to the formulations, calcium carbonates ($CaCO_3$), is extruded or pelletized into any shape. In terms of extrusion, different products can be used to produce supports with cylindrical geometries pierced by one to several channels: the barrels (Figure 2.8). The primary role of the central channel is to increase the macroporosity of the catalytic charge in order to facilitate the passage of large gas flows into the tubular reactors and to limit pressure losses. Its second role is to increase the S/V ratio (surface to volume), which allows the active surface of the catalyst to be increased. Finally, the reaction being highly endothermic, the geometry of the catalysts should also make it possible to enhance thermal transfer within the catalytic bed.

The cylinders or barrels obtained are then calcined and consolidated at a moderate temperature (typically 1,200°C), in order to limit the specific surface abatement, while providing the support with a mechanical resistance to compression. The ceramic supports thus obtained are then impregnated by being dipped into an aqueous solution of nickel precursors, such as nickel nitrates, for example. The catalysts obtained are finally calcined under air flow, in an inert or reductive atmosphere. Catalysts can be sold oxidized or activated (that is, after reduction of

$Ni^{II}O$ into Ni^0 under a reducing atmosphere). In this case, the reaction can be initiated as soon as the catalyst is loaded into the reactor. Catalysts can also be activated *in situ*. The process will then include a preliminary phase in which the temperature rises under a reducing atmosphere. Because methane is reductive, catalysts are often activated at temperature in the presence of methane.

Filling the reactors

Filling tubular reactors is a key step before starting a steam reforming unit. Usually, tubular reactors are randomly filled by catalysts. The most common method today is to use sheaths, which are long pockets of fabric that accompany the catalysts in their fall when filling the tubes. The volume of the tubular reactor to be filled is less than 0.1 m^3. It is possible to control the consistency of the catalysts from one tube to another by measuring pressure losses.

This value is known, for a given catalyst geometry (size, porosity), and must be reproduced in each of the tubular reactors with a tolerance of $\pm 3\%$. Poor filling of reactors can have dramatic consequences on the process, causing preferential flows in reactors presenting the lowest pressure drops, and exposing underfed tubes to the risk of overheating, which can lead to cracking.

2.1.2.3.4. The modeling of steam reforming furnaces

In order to maximize the productivity of a plant, the distribution of heat transmitted to each tube must be the most uniformly distributed. Numerical simulation helps to understand and optimize steam reforming furnaces, both in their design and operation. It takes into account the main physical and chemical phenomena involved in the combustion chamber and inside the reforming tubes.

In the combustion chamber, turbulent flames convert fuel and the oxidizer into heat. Fluid movements are described by the Navier-Stokes equations that account for the conservation of mass, momentum and energy, as well as the transport of the chemical species involved in the combustion reaction. Their resolution still represents a major technical challenge today.

Indeed, there is still no analytical solution to solve these equations by hand and it is necessary to program them by machine. Mixed skills are required at the interface between applied mathematics, physics and computer science. The execution of these programs requires significant computational resources: the numerical simulation of an industrial-scale steam reforming unit requires a supercomputer and a computational time of several days or even weeks.

The dominant mode of heat transfer inside a reforming furnace is radiation, at about 95%. The walls of the furnace exposed to high temperatures, as well as some combustion gases inside the furnace, such as carbon dioxide or steam, will contribute significantly to the radiative exchanges between the chamber and the reforming tubes. Radiation models must therefore be coupled with the Navier-Stokes equations to precisely describe the thermal transfer involved in the furnace. Depending on the problem, different simulation levels are possible.

Numerical simulation allows access to all of the physical information – temperature, composition, pressure, velocity fields – in any part of the combustion chamber. For example, Figure 2.9 shows *the structure of flames in high temperature areas*.

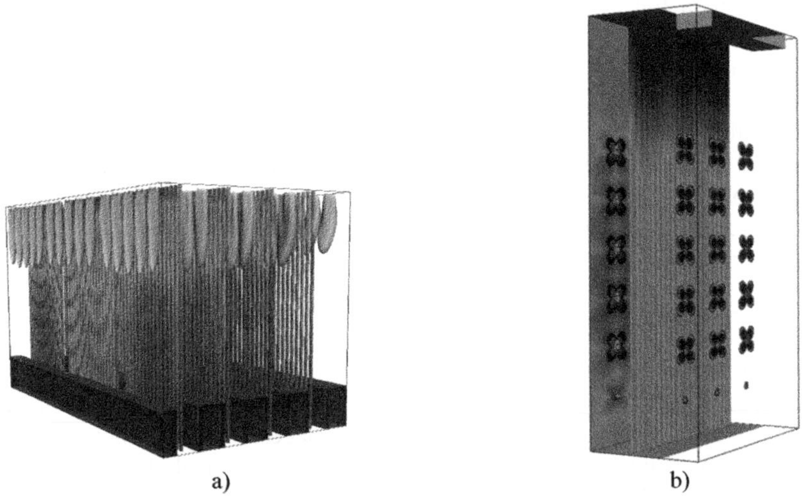

Figure 2.9. *Illustration of the results of simulating steam reforming furnaces: (a) top-fired; (b) side-fired. For a color version of this figure, see www.iste.co.uk/dalpont/process2.zip*

Inside the reforming tubes, models must describe the heat transfer from the outer surface of the tube to the inside of the tube. It is thus necessary to know the temperature of the gas in order to correctly estimate the synthesis gas production rate. The particularity of the reforming tubes is that the transfer of heat originating from the chamber is limited by the thermal resistance created between the tube and the pellets.

Chemical phenomena taking place inside a reforming tube (see section 2.1.2.3.1) are also taken into account through specific reaction models between species in the adsorbed phase. The parameters required for these kinetic models, such as pre-exponential factors, activation energies, adsorption enthalpies or thermodynamic equilibrium constants, are determined by experiments using catalysts crushed so finely that the internal diffusional limitation is negligible (Xu *et al.* 1989a).

However, at the industrial level, the diffusion of reactants in the pores of the pellets significantly limits the kinetics of the reaction, given the size and shape of the catalyst pellets. Effective reaction rate depends on the diffusion rates of the reactants towards the active sites (Xu *et al.* 1989b). The transport of reactants to the active sites of the catalyst pellets is limited by Knudsen diffusion, or more specifically by the collision effect between molecules and the surface of pellet pores, as well as intermolecular collisions. The corresponding diffusion limitation is modeled by efficiency factors that weight the reaction rate relative to the intrinsic kinetics.

At the scale of a full tube containing about ten thousand pellets, the computational methods restrict the modeling of these phenomena to a one-dimensional approach. More accurate three-dimensional simulations are possible in elementary domains containing a moderate number of combustion pellets (Figure 2.10).

Videos
Simulation of a steam reforming tube with a mean Reynolds simulation: https://youtu.be/O02VnFtpu0s

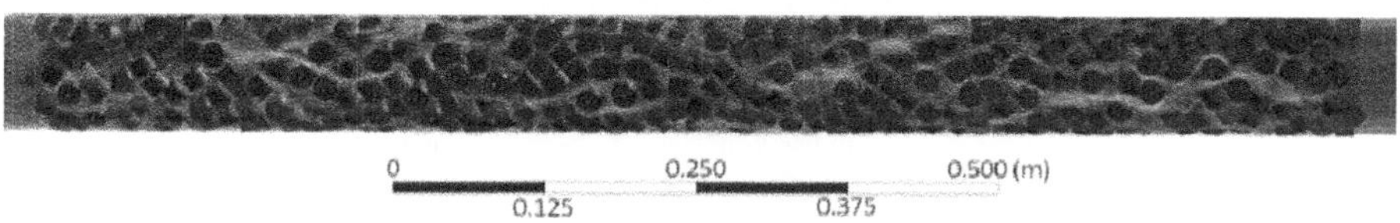

Figure 2.10. *Illustration of the three-dimensional simulation results for a section of tube. For a color version of this figure, see www.iste.co.uk/dalpont/process2.zip*

2.1.2.3.5. Heat transfer optimization inside the furnace

The design and construction constraints of furnaces can lead to an inhomogeneous transfer of heat to the reforming tubes, which is accompanied by a loss of efficiency. The main challenges to point out are the following:

– the definition of the design rules that determine the arrangement of the tubes and burners in rows and sections bounded by carrier beams;

– the design of a system for the distribution of air and fuel to the nozzles of the burners (Figure 2.11) as well as feed gas to the steam reforming tubes;

– to avoid the flame bending effects induced by impulsion differences between burners close to the walls and the others (Figure 2.12);

– the design of the fumes extraction system (Figures 2.13 and 2.14).

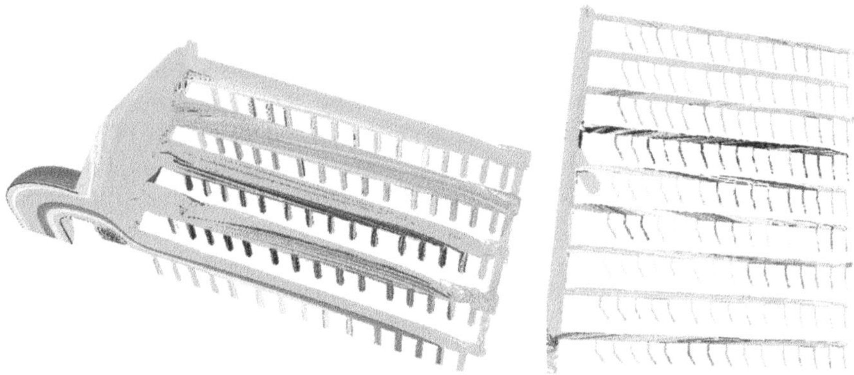

a) b)

Figure 2.11. *Illustration of the results of the three-dimensional simulation of flows in (a) an air distributor; (b) fuel. For a color version of this figure, see www.iste.co.uk/dalpont/process2.zip*

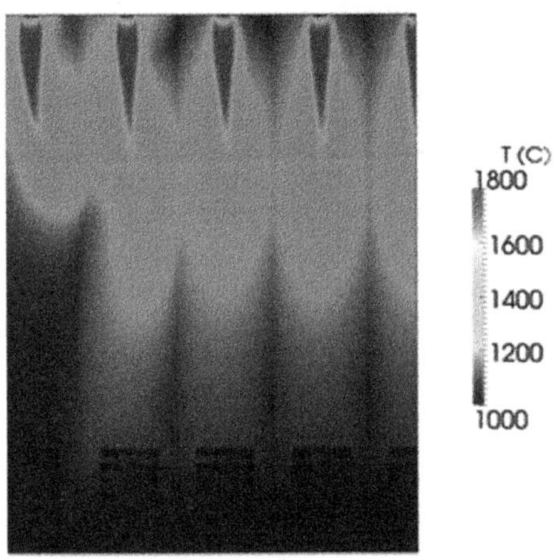

Figure 2.12. *Illustration of the flame bending phenomenon in a top-fired furnace. For a color version of this figure, see www.iste.co.uk/dalpont/process2.zip*

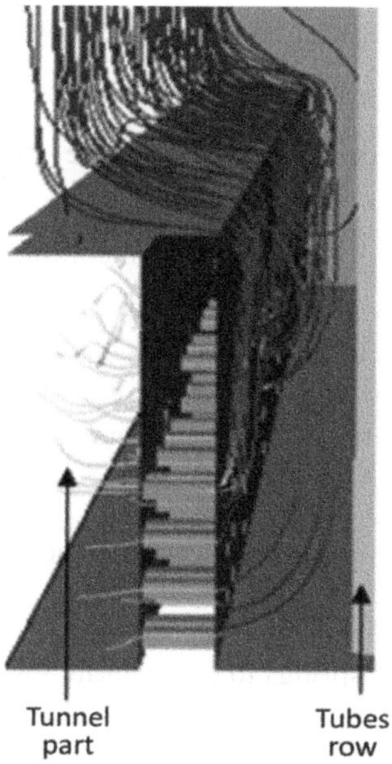

Figure 2.13. *Fumes flow in extraction tunnels. For a color version of this figure, see www.iste.co.uk/dalpont/process2.zip*

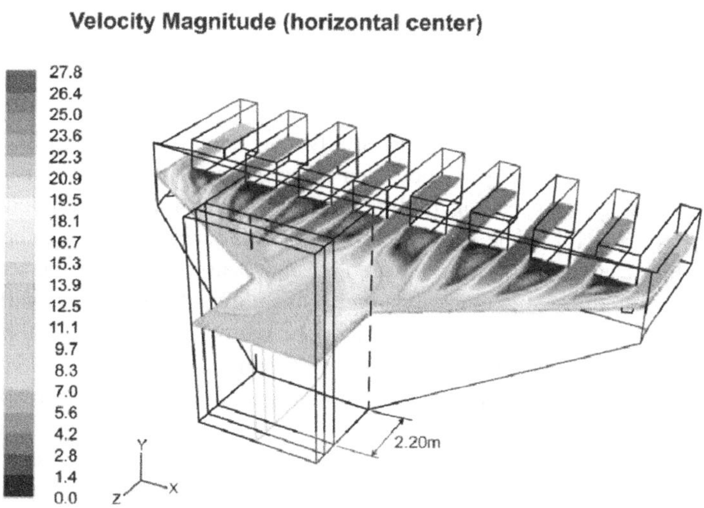

Figure 2.14. *Fumes flow in the converging pipe. For a color version of this figure, see www.iste.co.uk/dalpont/process2.zip*

2.1.2.3.6. SMR Reactants supply and product collection systems

The following are the different supplied gas and reformed gas distribution collectors:

– inlet collectors are generally made of type 316 austenitic stainless steel, including dilation lyres;

– collectors and outlet links are made of Incolloy 800 high-grade Cr-Ni alloy, due to temperature;

– large-diameter collectors, commonly known as transfer lines are in internally bricked steel. This refractory brick is made of silica-free alumina bricks (100% Al_2O_3). The carbon steel metal casing sometimes comprises a double envelope of cooling water or is cooled by water runoff.

Tube support is a key element in designing the furnace. Under the action of temperature variations, the tubes dilate or contract by several centimeters. It is therefore necessary to have semi-flexible links at the top and/or at the bottom, with counterweights or suspension springs to compensate for dilations.

2.1.2.4. *Hydrogen separation and purification*

2.1.2.4.1. CO conversion through the water gas shift reaction

At the reformer outlet, the synthesis gas contains between 10 mol% and 15 mol% of carbon monoxide, which can be converted into hydrogen by the water gas shift reaction. This reaction is carried out in a secondary reactor placed downstream of the steam reformer, after a boiler capable of cooling down the synthesis gas to the operating temperature of that reactor.

The water gas shift reaction is generally carried out in a fixed-bed adiabatic reactor (Figure 2.15) in the presence of a selective copper- or iron-based catalyst, depending on the type of reactor: high temperature (HT), medium temperature (MT) or low temperature (LT). The characteristics of these reactors are described in Table 2.3 and enable residual CO levels between 0.5 mol% and 5 mol% to be reached. It is interesting to note that a nickel catalyst can also catalyze the water gas shift reaction. However, it is not used here because of its poor selectivity: in fact, it also catalyzes the methanation reaction (reverse reaction of steam reforming: $CO + 3\ H_2 = CH_4 + H_2O$) and would lead to the production of methane in this secondary reactor.

Prior to the development and refinement of synthesis gas purification technology by adsorption (see section 2.1.2.4.2), purification was done by installing a high-temperature conversion reactor, followed by a low-temperature conversion reactor. Currently, the units are designed with a single conversion reactor at high or medium temperature. The choice of technology will depend on economic considerations regarding the selling price of the steam produced (see section 2.1.2.5.2).

Figure 2.15. *Water gas shift reactor*

	Low/medium temperature shift (LTS/MTS)	High temperature shift (HTS)
Catalyst	Cu ZnO	Fe Cr_2O_3
Inlet temperature	200–250°C	300–360°C
Outlet CO content	0.5–3 mol% dry gas	3–5 mol% dry gas
Steam/dry gas (S/DG) min ratio	LTS: 0.3/MTS: 0.22	No limitations

Table 2.3. *Comparison of the different possible configurations for a water gas shift reactor*

2.1.2.4.2. Hydrogen purification by adsorption (PSA)

Pressure Swing Adsorption (PSA) is the most common process for the industrial production of high-purity hydrogen, which can reach up to 99.999%. This process is based on the adsorption phenomenon, which is a phenomenon of interaction between molecules in a fluid phase and a solid adsorbent. The intensity of these interactions varies with the nature of the molecules and the solid adsorbent. For hydrogen purification by adsorption, molecules of impurities (nitrogen, carbon monoxide and dioxide, light hydrocarbons, etc.) exhibit stronger molecule/adsorbent interactions than hydrogen/adsorbent interactions. Since impurities are selectively retained (adsorbed) by the adsorbent, hydrogen can be purified. The higher the partial pressure of a species to be adsorbed in the gas phase, the more the solid adsorbent is charged with this species. Therefore, by varying the partial pressures of impurities in the gas phase, it is possible to vary the quantities of adsorbed impurities.

From a practical point of view, once the adsorbent is saturated with impurities, it must be regenerated (by desorption); this is why the process operates in cyclic mode. Regeneration is then done by lowering the pressure in the gas phase, so as to lower the partial pressures in impurities, and thus desorb them. Desorption of impurities results in the loss of some of the hydrogen in the tail gas. Generally different solid adsorbents are used in the form of successive layers, arranged in a fixed bed in the adsorbers (alumina, silica gels, active coals and zeoliths), each of these adsorbents stopping, respectively, water and ammonia, hydrocarbons and sulfur compounds, carbon dioxide, argon, oxygen and methane, and lastly, nitrogen and carbon monoxide.

This process has been industrially developed with a wide range of production capacity ranging from 50 Nm^3/h to more than 100,000 Nm^3/h of pure hydrogen for larger units. The cycles are different depending on the purity and recovery rate targets of hydrogen. They can make use of a number of adsorbers, typically ranging from 5 to 12 (see Figure 2.16).

Figure 2.16. *Principle flowsheet of a PSA-based hydrogen production unit. For a color version of this figure, see www.iste.co.uk/dalpont/process2.zip*

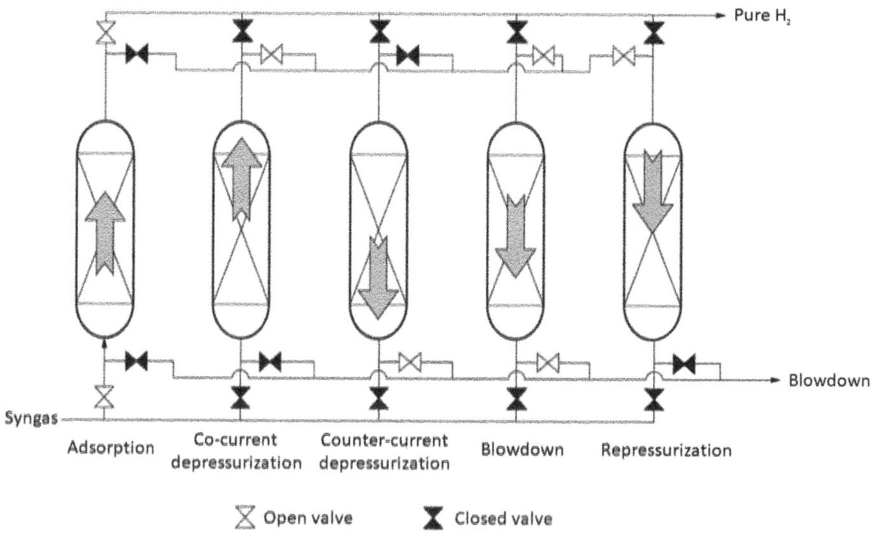

Figure 2.17. *Principle diagram of the operation of a PSA-based hydrogen purification unit*

Figure 2.17 illustrates the operating principle of a PSA, consisting of the following five basic stages:

– Stage 1 is adsorption: impure gas enters under pressure into an adsorber, impurities are adsorbed and pure hydrogen is produced. The gas usually flows upwards. When an adsorber has reached its adsorption limit, the gas to be treated is automatically switched to another regenerated adsorber;

– Stage 2 is co-current depressurization: in order to recover a significant part of the hydrogen present in the non-selective volumes of the adsorber, for example the interstitial volume, the adsorber is co-currently depressurized (in the same direction of flow as during the adsorption stage). The gas thus recovered (pure hydrogen) is used to purify and re-compress other adsorbers;

– Stage 3 is counter-current depressurization: at the end of the co-current depressurization step, the adsorber is partially regenerated by its counter-current depressurization (in the opposite flowing direction than that of the adsorption step) towards the blowdown, thus releasing impurities;

– Stage 4 is counter-current blowdown: the end of the regeneration is done by a counter-current gas sweep of the adsorber (in the opposite flowing direction than that of the adsorption step) with pure hydrogen from another adsorber in production, or an adsorber in the co-current depressurization phase;

– Stage 5 is repressurization: this step is achieved by pure hydrogen from the co-current depressurization of an adsorber and supplemented by a fraction of the hydrogen produced. Repressurization is usually done counter-current (in the opposite direction of flow to that of the adsorption stage) in order to concentrate pure hydrogen at the end of the adsorber, from which hydrogen is extracted during the next adsorption stage. Once the adsorber has reached the adsorption pressure, the cycle terminates and the adsorber is ready for a new adsorption stage (Stage 1).

As a general rule, the higher the number of adsorbers in service, the higher the recovery rate of the purified gas. Adding more adsorbers allows further splitting of the depressurization phases before blowing down the impurities, and increasing the recycling of depressurization gas as well as reducing the amount of blowdown.

Conversely, deactivating one or more adsorbers for maintenance purposes does not generally prevent the facility from operating, but the recovery rate of purified gas will be degraded.

The effect of adsorption pressure

The impurity content that the adsorbent can retain increases with pressure and, consequently, the processing capacity of the unit increases in conjunction with the pressure. The recovery rate also increases with pressure, though only moderately.

At lower pressure, the capacity of adsorption decreases, resulting in reduced cycle times and reduced recovery rates. At too low a pressure, the adsorbent of upper layers (dedicated to the most volatile compounds) can be saturated by heavy compounds not adsorbed by the lower layers. The adsorbent can then become non-regenerable.

The effect of desorption pressure

Since the adsorbent is regenerated by lowering the pressure, the lower the pressure, the more effective the regeneration will be. For a higher desorption pressure, the residual concentration of impurities on the adsorbent will be higher and the adsorption capacity in the next cycle will be reduced. The capacity of the unit will therefore be reduced for higher desorption pressure and increased for a lower desorption pressure.

The effect of pressure on the adsorption capacity

The working capacity is much higher with low desorption pressure than with a high desorption pressure. The lower the capacity of a work cycle, the more frequent the regenerations and the greater the quantity of gas blown down. The recovery rate of purified gas will then be lower. A PSA unit should always be operated with a blowdown pressure as low as possible. The effect of regeneration pressure will be as important on the capacity, as it is on recovery rate. This effect is represented in Figure 2.18.

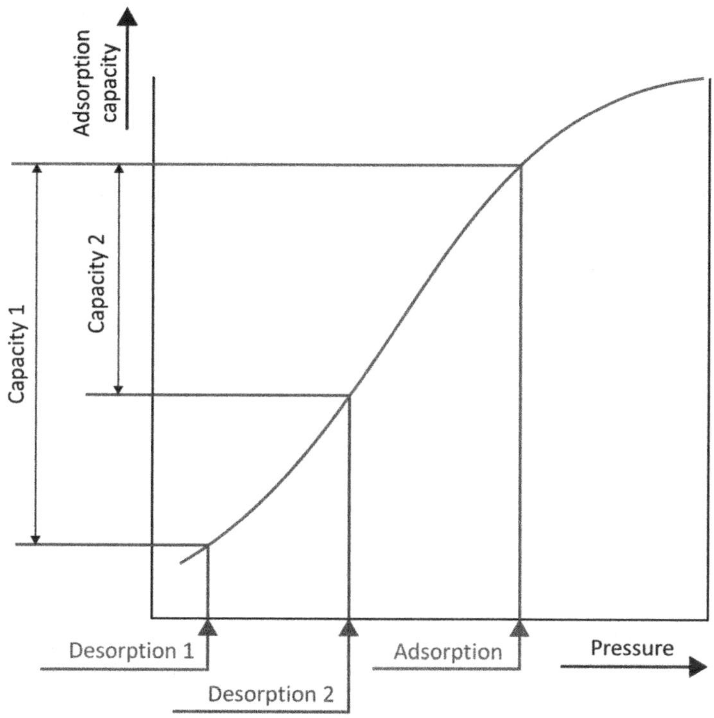

Figure 2.18. *The influence of adsorption pressure and desorption pressure on adsorption capability. For a color version of this figure, see www.iste.co.uk/dalpont/process2.zip*

The effect of temperature

The capacity of the adsorbent decreases as the temperature rises, however regeneration is more efficient at higher temperatures. Nonetheless, the shape of the adsorption isotherms (Figure 2.18) shows that the working capacity is lower at higher temperatures. The optimal operational temperature is between 20°C and 40°C.

The effect of the composition of the gas

Generally speaking, an elevated concentration of impurities in the feedstock gas reduces the capacity of the unit. On the other hand, if, due to an unplanned composition during the design of the unit, a heavy compound reaches a higher layer provided for light compounds, this compound will be perfectly and permanently adsorbed but poorly desorbed. It will permanently deactivate this adsorbent layer.

2.1.2.5. *Optimization of the process according to needs*

The design of a hydrogen gas production plant by steam reforming offers many possibilities to minimize operating costs. Depending on the needs of the end user, on the cost of raw materials and fuel, and on the recovery of the excess steam produced ("exported" steam) in relation to the need of the process, a certain plant configuration will be retained.

The key parameters guiding this choice are:

– the steam-to-carbon (S/C) ratio in the reformer feed;

– the implementation, or not, of a pre-reformer;

– the reformer inlet temperature;

– the reformer outlet temperature;

– the type of reactor implemented to convert carbon monoxide by means of the water gas shift reaction;

– the preheating, or not, of the combustion air;

– the reformer outlet pressure;

– the performance of the PSA purification unit.

For example, when the reformer feedstock is more expensive than the fuel used in burners, it is necessary to minimize feed consumption at the expense of fuel consumption.

This is possible by selecting the following configuration and operating parameters:

– a high steam-to-carbon (S/C) ratio in the reformer feed (> 3);

– a high reformer outlet temperature (> 900°C);

– two serial reactors implementing the water gas shift reaction, one at high temperature and the other at lower temperature;

– high PSA purification unit performance (89%).

2.1.2.5.1. Steam production

In the conventional steam reforming process, steam is produced by one or more boilers using the heat contained in the synthesis gas, at the outlet of the reformer, and the heat contained in the fumes generated by the furnace (see Figure 2.7).

This steam is generally produced in excess compared to the needs of the process itself and it can be sold to a customer, especially if the local situation allows it to be recovered at the energy value (Higher Heating Value, HHV) of the fuel used.

Depending on whether or not the valorization of excess steam is possible, it will be a matter of producing a maximum or minimum of excess steam compared to the needs of the process. This is possible by scrupulously choosing the configuration and operating parameters of the steam reforming process. It is understood that it is also possible to ensure excess steam production between these two limit values. Table 2.4 presents the main design differences between two plants, one designed to minimize excess steam production and the other designed to maximize it.

Steam reforming plant design	Excess steam	
	Minimum	**Maximum**
Combustion air preheating	Yes	No
Pre-reformer	Yes	Optional
Reformer feed temperature	High	Low

Table 2.4. *Differences in design between one steam reforming plant minimizing the excess of steam production and another maximizing it*

Although it is possible to significantly reduce the steam production of a hydrogen production plant, based on steam reforming technology in a multitubular furnace as shown in Table 2.4, it is not possible to economically design a plant with the same technology whose steam production in excess, compared to the need for the steam reforming reaction, would be zero. On the other hand, it is possible to obtain zero excess steam production with so-called convective reformer or reactor heat exchanger technologies, which make it possible to recover some of the heat from the synthesis gas at the reformer outlet to provide about 20% of the energy needed for the endothermic reforming reaction (see section 2.3.3).

2.1.2.5.2. The cost of hydrogen

The investment, the consumption of utilities (more precisely, energy consumption other than the energy contained in the unit feedstock: electricity, steam, fuel) and the energy efficiency of a steam reforming unit indeed depend on the purification method and the technology chosen. By integrating utility costs (positive and negative due to steam co-production), the cost price of hydrogen is approximately 1,700 euros/t assuming 12% return on investment and a depreciation period of 15 years. These costs do not include the capture costs of emitted CO_2, knowing that about 7 kg to 10 kg CO_2/kg H_2 are produced according to the production capacity, and the design retained.

2.2. Problems and operational constraints in steam reforming units

2.2.1. *Tube temperature and lifetime*

Reforming tubes have a limited lifetime due to irreversible deformations caused by the high temperatures and pressure they are subjected to. They are designed for a 100,000-hour operating period. The lifespan of the tubes depends on the temperature at which they are operated. Manufacturers guarantee a lifespan for each tube based on a maximal design operation temperature, known as DTT (Design Tube Temperature). Operating tubes at temperatures above the design temperature speeds up their aging and leads to premature deterioration (Figure 2.19). Operating 20°C above the design limit could reduce the life of the tubes by half. The accuracy of the temperature measurement in the tubes within the furnace is therefore essential for a safe unit operation.

Figure 2.19. *Tube failure in furnaces*

The most common method is the use of a pyrometer through optical peepholes, allowing visualization inside the furnace (Figure 2.20). If the pyrometer is not expensive in itself, the manpower for carrying out measurements is significant, especially for furnaces with hundreds of tubes in which each tube has to be measured one by one. In addition, the measurement of the tubes using an optical method is disrupted by the reflection of the surrounding walls, generally warmer than the targeted tubes.

Figure 2.20. *Tube temperature monitoring using pyrometers*

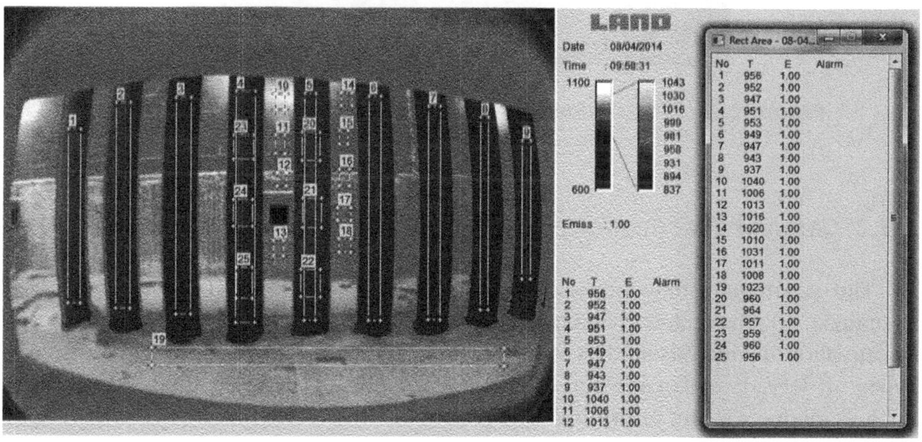

Figure 2.21. *Tube temperature monitoring using infrared cameras. For a color version of this figure, see www.iste.co.uk/dalpont/process2.zip*

An alternative that enables the measurement of the temperature of several tubes at once is the infrared borescope (Figure 2.21). Today this type of camera is very expensive, especially due to a specific wavelength, 3.9 μm, for which the impact of fumes on tube measurements is negligible. The price is also affected by regulations for equipment intended for use in ATEX zones (explosive atmospheres). In addition, the borescope requires water cooling and optical lenses to be cleaned with compressed air. The cooling and cleaning system increase the weight of the borescope, in particular compared to that of the pyrometer.

The measurement of tube temperature can also be performed by skin-type thermocouples associated with heat shields (Figure 2.22). In addition to the intrusive nature of this measure, another disadvantage is the limited number of areas where the temperature can be measured. On the other hand, the measurement is continuously performed and can be easily traced back to the unit control system.

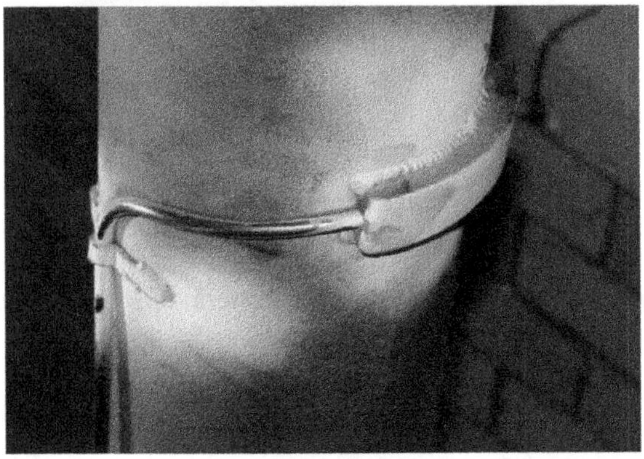

Figure 2.22. *Tube temperature monitoring using skin thermocouples equipped with a heat shield*

2.2.2. *Catalyst deactivation*

The steam reforming process may become less effective if the catalytic load contained in the tubes is deactivated. There are several causes of catalyst deactivation. Numerous studies focus on this subject. We consider here the different types of degradation of the active phase that could occur in an industrial process, such as methane steam reforming. The causes identified are:

– catalyst poisoning by chemical compounds;

– carbon formation;

– active phase degradation by friction;

– thermal degradations such as support or active phase sintering;

– active phase reaction with the support or with the gas phase.

The main causes of catalyst deactivation used in the steam reforming process are poisoning and coke formation. Poisoning is due to the presence of atoms or molecules that are irreversibly adsorbed at active sites. In the case of steam methane reforming, the nickel existing on the surface of industrial catalysts can be poisoned by sulfur, present in the natural gas. This poisoning is irreversible and leads to a change in the catalyst load. In order to prevent any risk of catalyst poisoning by sulfur, the natural gas is purified before it enters the main reformer (section 2.1.2.2.2). The use of a pre-reformer also helps in some cases to limit the poisoning of the catalytic bed in the main reformer.

With regard to carbon formation, the different reactions that could lead to it in a steam reformer are collected in Table 2.5.

Reactions		ΔH^0_{298} (kJ/mol)
1. Methane dissociation	CH_4 (g) ↔ C + 2 H_2 (g)	74.9
2. Boudouard equilibrium	2 CO (g) ↔ C + CO_2 (g)	– 172.4
3. Carbon reduction	H_2 (g) + CO (g) ↔ C + H_2O (g)	- 131
4. Hydrocarbon dissociation	C_nH_m → nC + (m/2) H_2 (g)	> 0
5. Hydrocarbon pyrolysis	C_nH_m → Olefins → coke	–
6. Gum formation	C_nH_m → $(CH_2)_n$ → gum	–

Table 2.5. *Reactions leading to carbon formation*

Reactions 1, 2 and 3 are equilibria, so the reverse reactions leading to carbon gasification can also occur under specific operating conditions, and for some compositions of the gaseous mixture. Reactions 4, 5 and 6 are total. To avoid them, a pre-reforming reactor can be placed before the steam reforming furnace (section 2.1.2.3). It allows for the conversion of heavy hydrocarbons into H_2 and CO, in order to avoid operating conditions in the furnace inlets that would lead to the formation of carbon by these reactions. Depending on the nature of the hydrocarbon feedstock treated by the steam reformer, the use of a pre-reformer is necessary or not.

The fight against carbon formation is a major challenge when operating a steam reforming hydrogen production unit. The operation parameters influencing the phenomenon, the temperature, the composition of the feeds to be treated and the S/C ratio (molar ratio between steam and carbon atoms entering the process) are controlled. Carbon formation in a unit is therefore the result of an abnormal operation of the unit, for example:

– a steam defect relative to the hydrocarbon feed to be treated;

– operating conditions too close to the thermodynamic limits of carbon formation (operation at too a high temperature, etc.);

– catalyst poisoning.

Three carbons of different natures can be produced as part of an abnormal steam reformer operation. Their production conditions vary according to composition, temperature and pressure, and, consequently, depending on the location in the overall steam reforming process. The formation mechanisms of these three types of carbon are collected in Table 2.6.

Carbon type	Filamentary	Gums	Pyrolytic
Temperature range	> 450°C	< 500°C	> 600°C
Formation mechanism	Dissociation of hydrocarbon and CO on the surface of the catalyst (Ni) leading to the nucleation and then growth of a carbon filament	Polymerization of hydrocarbon radicals on the surface of the catalyst (Ni) leading to the encapsulation of the active particle by a carbon film	Thermal cracking of hydrocarbons in gas phase (non-catalytic)
Impact	No catalyst deactivation, the active particle being supported by the carbon filament. The growth of the filaments leads to the bursting of the catalyst and to the increase in pressure losses in tubes, as well as to the formation of hot spots	Gradual catalyst deactivation by progressively blocking the active surfaces of the catalyst	Carbon deposit in the catalytic bed resulting in loss of activity and increased pressure loss
Critical parameters	Low S/C, temperature too high, presence of olefins, of aromatics	Low S/C, lack of hydrogen, low temperature, presence of aromatic compounds	High temperature, presence of olefins, sulfur poisoning
Process areas concerned	Pre-reformer: surface dissociation of CH_4 catalyst or C_nH_m Steam reformer tube: cracking on the surface of the CH_4 catalyst, Boudouard equilibrium	Pre-reformer: surface cracking of C_nH_m followed by polymerization	Pre-reformer: C_nH_m gas phase pyrolysis

Table 2.6. *The different types of carbon that can be produced by a steam reforming process*

Presence of water limits the risk of carbon formation at high temperatures. One solution to prevent this is to increase the S/C ratio to ensure that carbon cannot be thermodynamically formed under these conditions.

This solution has the advantage of producing a synthesis gas rich in hydrogen. However, it is expensive in terms of energy because the water needs to be preheated before being injected with the other reactants into the reformer.

When the carbon formation reaction is catalytic, it occurs at the acidic sites of the active phase support. This process can be stimulated by the presence of Ni.

Additives can also be added to the formulation of the catalyst support or when the active phases are impregnated, mainly alkaline metals, such as potassium, which alters the acid-basicity of the support and prevents the formation of coke.

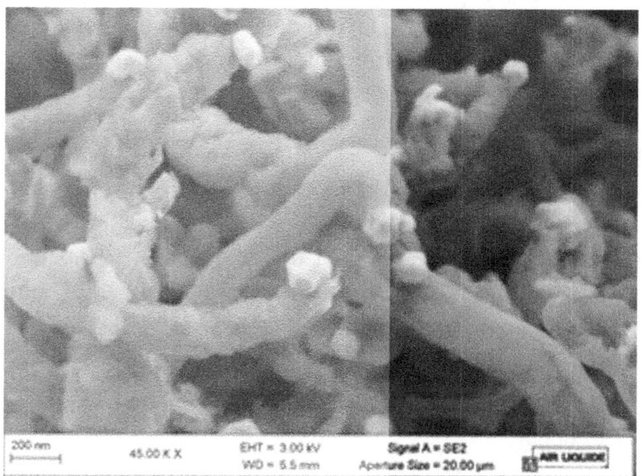

Figure 2.23. *Micrograph MEB showing the presence of carbon filaments on the surface of an SMR catalyst*

The deactivation by friction of catalytic pellets between one another is quite limited in this process because the catalytic bed is fixed and the reactants usually cross the bed from top to bottom, to minimize the risk of catalyst fluidization.

The catalytic supports of the active phase chosen for this reaction – magnesium- or calcium-doped alumina – are relatively thermally and chemically inert. However, there has been a significant decrease in the specific surface area of the support after use. The specific surface abatement reflects a deconstruction of the support surface by sintering (loss of porosity, increase in grain size). Ni particles existing on the support can get trapped in the support during sintering and thus be inaccessible to the reactive gas mixture. This deactivation mechanism can help reduce Ni dispersion after use.

Finally, methane steam reforming catalysts can also become deactivated by coalescence of Ni particles that constitute the active phase. The accessibility and size of the metal sites of commercial catalysts have been characterized. The metal surface is directly correlated to the number of active sites, thus to the efficiency of the catalyst. Metal dispersion represents the amount of metal accessible to gas, relative to the total amount of metal.

2.2.3. *Corrosion by metal dusting*

In all steam reforming processes involving heat exchangers in which high-temperature processed gas circulates, avoiding corrosion by metal dusting represents a challenge. Metal dusting is a phenomenon of carburization that can occur on metal surfaces under specific conditions, especially at temperatures between 400°C and 800°C and in the presence of a gas with high-carbon activity. This phenomenon, although well known in the industry, is unfortunately not yet fully understood, yet many studies are regularly published to analyze the different mechanisms involved according to the types of alloys and the nature of the metal surface. The main effect of corrosion by metal dusting is a loss of material and the formation of metal dust containing not only metal but also metallic carbides and carbon. This phenomenon can be avoided by:

– adding sulfur to the processed gas, but in the case of steam reforming, sulfur has been removed from the feed for other reasons upstream of hot areas where metal dusting is likely to occur;

– choosing appropriate building materials;

– applying protective coatings – deposits of oxide layers (aluminizing) – to the materials used.

The equipment can also be designed and sized so as to avoid the operating area in which this phenomenon can occur. High-temperature surface or steam treatments can significantly increase the time it takes to develop; on the other hand, other treatments conventionally used such as pickling will have the opposite effect.

2.2.4. *Flexibility in raw materials of steam reforming units*

Traditionally, feedstocks used in steam reforming units are natural gas, naphtha and liquefied petroleum gases (LPG). Nevertheless, the current trend consists of designing hydrogen production units by steam reforming capable of using different raw materials originating from the refining industry, such as residual gases rich in hydrogen from refineries or high-concentration olefin feeds. This growing demand for flexibility in units leads to the development of new designs and new catalysts. One of the key points in the design of this type of unit is the pre-reformer. Air Liquide operates several units capable of processing different raw materials (Figure 2.24, section 2.3.2).

Designing and sizing a hydrogen plant capable of operating with different raw materials makes it possible to use the most economical material at all times.

Location	**Saudi Arabia**	**United States**	**Europe**
Unit type	H2 Plant	H2 Plant	H2 Plant
Raw material	Natural gas	Natural gas	Natural gas
	Residual refinery gases	Butane	LPG
	Propane	Pentane	Naphtha
	Butane	Ammonia	-
	Naphtha	-	-

Table 2.7. *Examples of units managed by Air Liquide, operating with different kinds of feedstocks*

2.3. Recent industrial developments responding to global warming

2.3.1. *The role of hydrogen in energy transition (Hydrogen Council 2017)*

Hydrogen is a key molecule in energy transition that is essential to limit global warming to 2°C. Hydrogen can play seven major roles in this transition, requiring a drastic reduction in CO_2 emissions:

– enable the large-scale integration of renewable energy by converting and storing renewable electricity produced in excess compared to demand;

– distribute energy between regions and sectors by connecting regions rich in renewable energy and other importing energy, at an international level;

– act as a buffer to improve the resilience of energy systems by storing electricity and becoming a strategic reserve;

– decarbonize transport;

– decarbonize industrial energy utilization, for example, for medium- and high-temperature processes, for which the use of electricity cannot be considered;

– help decarbonize the use of heat and electricity in buildings;

– provide a clean raw material for the industry.

In light of the critical role that hydrogen played in the energy transition, it is essential that new technologies be developed for producing hydrogen that emit less CO_2. The following chapters present developments made in recent years, within the steam reforming process, in order to reduce its carbon footprint.

2.3.2. CO_2 capture in hydrogen production units

The global environmental debate could lead to the need for steam reforming hydrogen production plants from fossil fuels to include carbon dioxide sequestration. The production of 1 kg of hydrogen by natural gas steam reforming also produces 9 kg of CO_2. This is the reason why Air Liquide has developed a technology that can capture the CO_2 released during hydrogen production. This technology – Cryocap^{TM} H_2 – not only makes it possible to reduce CO_2 emissions during the production process, but also significantly increases the productivity of the hydrogen plant.

The Cryocap^{TM} H_2 process separates and purifies the carbon dioxide, contained in the waste gas of the PSA unit of a hydrogen production unit, by steam reforming natural gas. This is a cryogenic method of capturing carbon dioxide. A the simplified diagram of the operation of a steam reforming unit, equipped with a Cryocap^{TM} H_2 unit, is summarized in Figure 2.24.

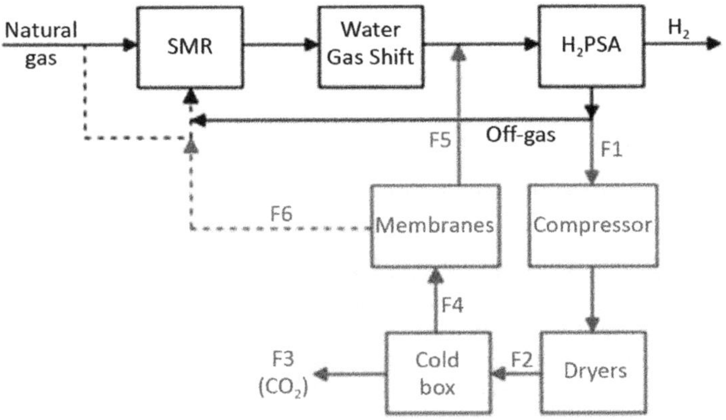

Figure 2.24. *Diagram of a steam reforming unit (in black) equipped with a Cryocap^{TM} H_2 unit (in red). For a color version of this figure, see www.iste.co.uk/dalpont/process2.zip*

COMMENT ON FIGURE 2.24.– *F1: fraction of PSA waste gas sent to the Cryocap™ H_2 unit. F2: stream exiting the dryers. F3: CO_2 stream produced. F4: non-condensable gas stream exiting the cold box. F5: H_2-rich stream returned to the SMR unit. F6: waste gas sent to burners.*

As shown in Figure 2.24, the four main steps of the Cryocap™ H_2 process are: compression and drying of the feeding flow F1, the cryogenic carbon dioxide separation unit, and the membrane separation that allows the recycling of a stream rich in hydrogen, within the process. The cryogenic unit separates CO_2 from other components mainly through the combination of partial condensation and distillation operations.

Natural gas is composed mainly of methane (CH_4), but it also contains very small amounts of nitrogen and light hydrocarbons, such as ethane (C_2H_6), ethylene (C_2H_4) and propane (C_3H_8). Natural gas steam reforming allows a partial conversion of methane, and a near total one for other hydrocarbons, into carbon monoxide (CO) and hydrogen. The synthesis gas produced after cooling feeds the water gas shift (WGS) reactor in which the WGS reaction enables the production of hydrogen and carbon dioxide from carbon monoxide and water.

The gas rich in hydrogen thus produced is then sent to the purification unit (PSA) in which the hydrogen is separated from the residual gas that contains, in addition to hydrogen, all the other components initially present in the hydrogen-rich gas.

In the reactor implementing the WGS reaction, in addition to this main reaction, secondary Fischer-Tropsch reactions occur and lead to the formation of saturated and unsaturated hydrocarbons, as well as methanol, in the form of impurities that will be found in the residue gas that leaves the hydrogen PSA.

The stream F1 contains about 46 mol% of CO_2, 22 mol% of H_2, 13 mol% of CO, 17 mol% of CH_4, 0.4 mol% of water, in addition to the nitrogen that is contained in the natural gas that feeds the steam reforming reactor, traces of light hydrocarbons (C_2H_6, C_2H_4, C_3H_8, C_3H_6, and benzene) as well as traces of methanol (CH_3OH) and ammonia (NH_3) which are co-produced from the water gas shift reaction.

The F1 stream is compressed and then sent to dryers. When the F2 stream leaves the drying process, it has a composition similar to that of the F1 stream, except for water and ammonia because these components are completely eliminated due to temperature modulation adsorption in the dryers.

The cold box works with temperature values of around -55°C and constitute the core of the Cryocap™ H_2 solution. The synergy between the different operations that take place inside the cold box (in particular, the first separator pot and multi-staged distillation) and then the catalytic oxidation in the presence of oxygen ensures the separation of CO_2 from the other components. The F3 stream coming out of this section of the process is made up almost exclusively of CO_2.

Non-condensable compounds are collected from the cold box and recovered by additional hydrogen production through membrane modules. In fact, the F4 stream feeds two membranes in series that can separate an F5 stream rich in H_2, recycled upstream of the hydrogen PSA, from a residual gas that, given its methane molar fraction, is mixed with the natural gas destined for combustion, producing the heat required by the steam reforming reaction.

This method makes it possible to obtain food grade carbon dioxide.

The main advantages of this technology are:

– a reduced ecological footprint by reducing CO_2 emissions by 60%. This solution represents an important step towards the eventual production of decarbonized hydrogen;

– an increase in hydrogen production of 10% to 20%;

– the lowest cost on the market for CO_2 capture in hydrogen production units (especially compared to the activated MDEA scrubbing process);

– a technology adapted to existing and future hydrogen production units.

This solution does not really compete with one of the current configurations which consists of capturing CO_2 in the gas upstream of the PSA with an amine scrubbing unit (MDEA) and recycling the CO_2 recovered in the reformer in which it will react with methane according to the dry reforming reaction:

$$CH_4 + CO_2 = 2CO + 2H_2, \Delta H_{25°C} = 247 \text{ kJ/kmol} \quad [2.4]$$

Furthermore, CO_2 recycling is mainly carried out in units that co-produce hydrogen and carbon monoxide; because in these units, CO_2 capture is imperative upstream of the cryogenic distillation unit of carbon monoxide to avoid solid CO_2 formation at very low temperatures. In addition, in these units, recycling CO_2 allows the CO/H_2 ratio produced to be increased, which is not desirable in units producing hydrogen only.

2.3.3. *The exchanger reactor ("zero steam")*

The production of hydrogen is associated with CO_2 emissions. In the steam reforming process, CO_2 emissions come from two distinct sources: CO_2 is produced both by the water gas shift reaction and by combustion in the reformer burners. Given that these two sources have the same relevance, reducing the amount of energy brought by combustion reduces CO_2 emissions.

To this end, several exchanger reactor concepts have been developed by different companies, using a reforming tube technology called "tube-in-tube". The idea behind this concept is simple: after conversion within the catalytic bed, the hot reformed gas enters one or more internal tubes arranged counter-currently relative to the catalytic bed (Figure 2.25) (Schlichting *et al.* 2016). All of these elements together constitute the reactor. As such, in this configuration, the energy contained in the reformed gas is partially used to provide about 20% of the energy needed for the endothermic reforming reaction. On the one hand, this reduces the amount of fuel sent to the burners and therefore the carbon dioxide emissions but also, on the other hand, the heat available for steam generation in the boiler installed downstream of the reactor and thus the steam production of the process waste steam. This makes it possible to obtain an efficient "zero-steam" hydrogen production unit, that is not producing excess steam.

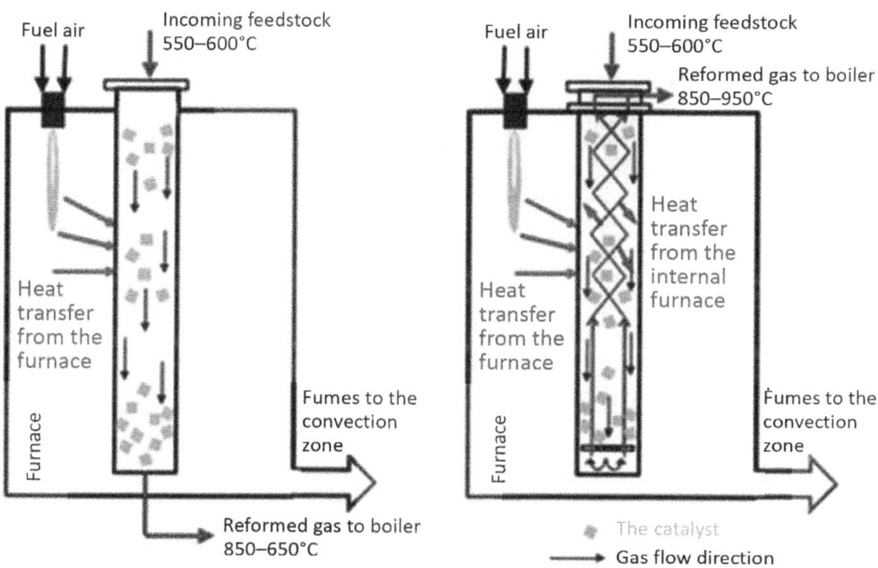

Figure 2.25. *(a) Conventional steam reforming furnace; (b) furnace using "tube-in-tube" technology. For a color version of this figure, see www.iste.co.uk/dalpont/process2.zip*

One of the major challenges with this technology consists of developing a material solution for internal tubes that can withstand corrosion due to high temperature metal dusting (see section 2.3). Several technologies making use of the "tube in tube" concept are currently on the market: Topsoe Bayonet Reformer™ (TBR) and Haldor Topsoe Convection Reformer™ (HTCR) by Haldor Topsoe, SMR-X™ by Air Liquide.

In order to develop this new technology, Air Liquide has invested an industrial-sized demonstration unit, named Callisto (Figure 2.26), capable of performing long-term testing. Following the success of the demonstration, this new generation of steam reformer was put on the market in 2018. Internal heat recovery results in a reduction of 5% of the natural gas consumption, as well as CO_2 emissions. The first commercial unit is scheduled for use in 2020.

Figure 2.26. *Callisto demonstration unit of tube-in-tube technology*

2.3.4. *Current research interests*

2.3.4.1. *Process intensification*

Today, many industrial sectors face challenges such as scarcity of resources, reduction of environmental footprint, increasing demand accompanied by increased international competition. Production facilities must therefore constantly innovate to maintain the highest levels of energy efficiency, productivity and safety. It is within this context that the concept of process intensification has emerged, which proposes an approach to respond to these challenges, using so-called disruptive technologies or methods. It is therefore part of a sustainable development approach.

Process intensification is often confused with continuous process improvement; nevertheless, it differs significantly from the latter by a desire to completely rethink existing processes, as evidenced by the many definitions that emerged during the

1990s and 2000s. The first definition of process intensification was proposed by Colin Ramshaw in 1995 at the opening of the first international conference on process intensification in the chemical industry:

> A strategy for making dramatic reductions in the size of a chemical plant so as to reach a given production objective.

This definition has been repeated many times and expanded to converge in a new definition proposed by Stankiewicz and Moulijn in 2000: "Any chemical engineering development that leads to a substantially **smaller, cleaner and more energy efficient technology**."

Finally, in 2007, a survey was conducted within various industrial sectors and resulted in a publication entitled European Roadmap for Process Intensification in which the following definition was given:

> Process intensification (PI) presents a set of often radically innovative principles ('paradigm shift') in process and equipment design, which can bring significant benefits in terms of process and chain efficiency, capital and operating expenses, quality wastes, process safety and more.

Process intensification thus proposes a method or technology adapted to a physical or chemical phenomenon. It is based on a few fundamental principles aimed at optimizing chemical transformation by being as close as possible to intra- and intermolecular events:

– ensure the same treatment for each molecule;

– reduce limitations due to transfers;

– find the effects of synergy between phenomena and/or operations.

To achieve these goals, a multiphysical and multi-scale approach is then envisaged, based on four founding pillars:

– structure (the spatial scale);

– thermodynamics (the operating window and energy supply);

– synergy (multifunctionality and/or hybridization);

– time (the time scale).

In the face of these solutions, it is necessary to assess the limitations of the process studied, in order to identify the most effective solution offered by process intensification.

2.3.4.2. *Limitations of the steam reforming process*

The performance of the conventional steam reforming process is limited by:

– intraparticular diffusion in the pores of the catalyst: the steam reforming reaction is not only controlled by reaction kinetics but also by the mass transfer from reactants towards active sites. Consequently, 95% of the catalyst charged in a reactor does not actually catalyze the reaction (De Deken *et al.* 1982; Soliman *et al.* 1988);

– heat transfer: the efficiency of the steam reforming furnace is controlled by thermal transfer 1) by radiation and convection from flames and fumes to the outer wall of the tube, 2) by conduction in the tube wall, 3) by radiation, convection and conduction from the inner wall of the tube to the gas and catalyst;

– the constraint imposed by the thermodynamic equilibrium of the reforming reaction: the composition of the synthesis gas at the reactor outlet is generally close to the composition at equilibrium, the latter being dependent on the reforming temperature, the operating pressure and the steam-to-carbon ratio (S/C) in the feed.

2.3.4.3. *Process intensification applied to steam reforming: the microstructured reactor (additive manufacturing)*

A technological solution provided by process intensification which can reduce the limitations related to heat and mass transfer is the use of **a micro or millistructured reactor with a catalyst deposited on a thin wall layer** to achieve the steam reforming reaction. This technological solution is based on a **structural approach** and aims to minimize limitations by reducing the characteristic dimensions.

Thus, as the catalyst is deposited on a thin layer of the wall (typically a few microns), the characteristic dimension associated with the transfer of internal mass is greatly reduced compared to traditional technologies and this limitation can be lifted.

Similarly, hydraulic passages whose order of magnitude is between one micron and one millimeter, contribute to a decrease in the characteristic dimension associated with heat transfer. In addition, the structure of the equipment offers very large surface-to-volume ratios; the heat transfer limitation of the conventional process can also be lifted. Therefore, this technology seems to offer an intensified solution relevant to the steam reforming reaction.

Moreover, as explained in section 2.1.2.3, to bring the heat needed to achieve the steam reforming reaction, the transformation is conventionally carried out in a furnace. Depending on the operating conditions and the design of the reformer, between 45% and 55% of the heat released by combustion into the burners is transferred to the reforming tubes, the remaining heat being recovered in the convection area to achieve different preheatings (see section 2.2.2.3.2). In order to obtain a highly efficient process overall, steam reforming uses strong thermal integration, in particular with heat recovery from these fumes and products, to preheat process fluids and also to produce steam, generally in excess compared to the needs of the process, as explained in section 2.2.2.5.1. For an on site hydrogen production unit (25–500 Nm^3/h H_2), it is not possible to export steam and its decrease leads immediately to an increase in the efficiency of the process. This approach is implemented as part of the tube-in-tube technology described in section 2.3.3. The reactor used is therefore a reactor exchanger, which is an intensified piece of equipment, based on the development of synergy between two conventional unit operations of process engineering, a reactor and an exchanger.

Finally, it appears that a particularly interesting technology offered by process intensification for steam reforming can be a micro or millistructured reactor exchanger with a catalyst deposited on a thin wall layer.

For example, since the late 1990s and early 2000s, many academics and industrialists have been focusing on the application of these technologies for steam reforming process, making it possible to generate proofs of concepts at pilot scale, as well as that of industrial demonstration units. Since the 2000s, Air Liquide has been promoting research and development projects for the use of such technologies for the steam reforming process, paying attention to the relationship between manufacturing processes and design.

Early developments were inspired by plate-fin heat exchangers, which typically contain hundreds of channels, formed by alternating plates and fins welded together. Producing them, either by brazing or by diffusion bonding, has been very difficult. There is still no equipment in refractory alloy on the market today, such as Ni-based super alloys, which can withstand the operating conditions of the steam reforming process (refractory alloys are required because the materials are subjected to creeping behavior). Moreover, these traditional methods of development restrict geometric possibilities for this equipment, limiting the potential for intensification. It is within this context that additive manufacturing (AM) can be used today, to free the design of this equipment and thus make it more efficient, without sacrificing the cost of manufacturing, which must remain competitive.

As early as 2014, Air Liquide experimented with the manufacture of the first reactors using additive manufacturing, initially on the basis of the same design used for the manufacturing of prototype reactors by brazing, and then by adapting the design to new opportunities offered by this innovative means of production. The first prototype produced was already 30 cm long and had a cross-section of 100 cm^2 (Figure 2.27).

Figure 2.27. *The first reactor produced using AM in 2014*

This first reactor was installed in a pilot plant producing hydrogen by SMR, installed within Air Liquide's Innovation Campus in Paris as early as 2015, and successfully tested less than six months after the manufacturing plans were sent for production, thus demonstrating the great benefits of AM for prototyping. The reactor was operated for more than 5,000 hours, at temperatures of up to 900°C and pressures of more than 10 bar, making this prototype the first pressure equipment achieved by additive manufacturing. Very promising results have been brought forward, demonstrating that MRS can be intensified by reducing the size of the reactor down to millimeter channels. These results have recently been awarded by the European Federation of Chemical Engineering (EFCE) and this has led to them appearing in various publications.

The main advantage of AM is to free up design rules that are constrained by manufacturing processes. As such, the internal components of the reactor can now be thought of in 3D, which was totally impossible with conventional manufacturing processes. The new possibilities offered by AM have been used to improve heat and mass transfer properties in the reactor by:

– creating new internal structures that can improve fluid distribution and mixing properties, especially for collectors and distributors;

– creating new shapes for the envelopes of pressure resistant reactors, but with smaller sizes/thicknesses, especially for collectors and the reactor body;

– proposing a new design for channels, providing heat more efficiently and making it possible to use a very small amount of catalyst.

For example, the distributors of microstructured reactors have specific shapes, which can only be manufactured by AM, to improve fluid distribution between each channel.

To design reactors such as these, heat transfer and reaction kinetics must be taken into account; the complete design must also optimize the distribution of gas flow between channels and must be resistant enough for applications in high-temperature and high-pressure processes. There is currently no software commercially available to simulate this type of operation. The design of these reactors is made possible by the development of a simplified 1D model, capable of processing thermal exchanges and reactions as they occur in the reactor. Moreover, the design of these reactors combines CFD and thermomechanical computations, in order to make the most of this revolutionary new manufacturing method, for example by minimizing material thicknesses. These tools can be used to scale equipment and predict performance.

Figure 2.28. *Examples of reactors produced by AM since 2014: evolution of dimensions and shapes. For a color version of this figure, see www.iste.co.uk/dalpont/process2.zip*

This technology looks particularly promising in the case of a low-capacity hydrogen production unit (in the range of 25 to 500 Nm^3/h of H_2) installed on site.

2.3.5. *Other means to provide reaction heat (currently in development)*

Currently, other solutions are being explored to provide reaction heat while minimizing CO_2 production.

One of the most advanced is the supply of heat by electric heating, either by Joule effect or through the generation of an electric field. It is also worth noting that there is ongoing research and development work on microwave and induction heating, as well as ultrasound heating.

2.4. References

De Deken, J., Devos, E.F., Froment, G.F. (1982). Steam Reforming of Natural Gas: Intrinsic Kinetics, Diffusional Influence and Reactor Design. In *Chemical Reaction Engineering*, Wei, J., Georgakis, C. (eds). American Chemical Society, Boston.

EFCE (2017). EFCE Process Intensification Award 2017, Press release [Online]. Available at: https://efce.info/efce_media/-p-3380-EGOTEC-10d31beae69d5159e93910ac1de4f484.pdf.

Giroudière, F., Le Gall, A. (2016). Production des gaz de synthèse par vaporeformage [Online]. Available at: https://www.techniques-ingenieur.fr/base-documentaire/procedes-chimie-bio-agro-th2/procedes-industriels-de-base-en-chimie-et-petrochimie-42329210/production-des-gaz-de-synthese-par-vaporeformage-j5480/.

Gourdon, C. (2016). Intensification des procédés – Fondamentaux et exemples d'industrialisation [Online]. Available at: https://www.techniques-ingenieur.fr/base-documentaire/procedes-chimie-bio-agro-th2/innovations-en-genie-des-procedes-42487210/intensification-des-procedes-j7002/.

Guisnet, M., Naccache, C. (2004). Catalyse hétérogène – Mode d'action des catalyseurs [Online]. Available at: https://www.techniques-ingenieur.fr/base-documentaire/procedes-chimie-bio-agro-th2/catalyse-et-procedes-catalytiques-42325210/catalyse-heterogene-j1250/.

Hydrogen Council (2017). Hydrogen Scaling Up, Report [Online]. Available at: http://hydrogencouncil.com/wp-content/uploads/2017/11/Hydrogen-Scaling-up_Hydrogen-Council_2017.compressed.pdf.

Mbodji, M. (2013). Conception et dimensionnement de réacteurs-échangeurs microstructurés pour la production de gaz de synthèse par vaporeformage du méthane. PhD thesis, University of Lorraine, Metz.

Mbodji, M., Commenge, J.M., Falk, L. *et al.* (2012). Steam methane reforming reaction process intensification by using a millistructured reactor: Experimental setup and model validation for global kinetic reaction rate estimation. *Chemical Engineering Journal*, 207–208, 871–884.

Plehiers, P.M., Froment, G.F. (1990). Coupled simulation of heat transfer and reaction in a steam reforming furnace. *Chem. Eng. Technol.*, 12, 20–26.

Ramshaw, C. (1995). The incentive for process intensification. Proceedings, *1st intl. Conf. Proc. Intensif. for Chem. Ind.* BHR Group, London.

Rostrup-Nielsen, J.R. (1983). *Catalytic Steam Reforming*. Springer-Verlag, Berlin.

Schlichting, H., Ulber, D., Cadalen, S., Doublet, S., Prost, L. (2016). High efficiency zero export steam reforming. *Petroleum Technology Quarterly*, 3, 99–105.

Soliman, M.A., Elnashaie, S.S.E.H., Al-Ubaid, A.S., Adris, A.M. (1988). Simulation of Steam Reformers for Methane. *Chemical Engineering Science*, 43, 1803–1809.

Stankiewicz, A.I., Moulijn, J.A. (2000). Process intensification: Transforming Chemical Engineering. *Chemical Engineering progress*, 22–34.

The Chemical Engineer (2018). Add it up! [Online]. Available at: https://www.thechemicalengineer.com/features/add-it-up/.

Xu, J., Froment, G.F. (1989a). Methane steam reforming I. Methanation and water-gas shift: Intrinsic kinetics. *AIChe Journal*, 35(1).

Xu, J., Froment, G.F. (1989b). Methane steam reforming II. Diffusional limitations and reactor simulation. *AIChe Journal*, 35(1).

3

Industrialization: From Research to Final Product

PRELIMINARY NOTE.– This chapter covers the industrially very important areas of process development, project management and industrialization including engineering, which is its implementation phase. In fact, it barely covers them; a full study would require several books. We tried to highlight key concepts around references that should allow readers to effectively engage with experts and simply find their way. For example, the notions of project owner and general contractor or prime contractor, which often prove confusing.

Regulatory aspects such as building permits are merely mentioned. The same is true of return on investment.

The reader will find in the books referenced (Charpentier *et al.* 2014; Dal Pont 2012; Dal Pont and Azzaro-Pantel 2014; Poux 2010) additional information regarding the notions developed hereafter.

The purpose of the industrial process enterprise is to bring to market use-value products, as explained in Chapter 1 of the first volume. **The final goal of the company is to sell**. To this end, it implements means of production (workshops, plants) that are specific to the company, which it shares with third parties in partnerships or JVs (joint ventures) or that it sub-contracts to specialized companies in whole or in part.

Imagine a drug of a complex chemical nature resulting from the transformation of multiple raw materials that undergo successive reactions, separations, or purification. At the end of the chain this drug will be packaged and shipped (see

---

Chapter written by Jean-Pierre DAL PONT.

Chapter 2 of the first volume). These steps can be carried out in the same workshop (unit), in different workshops of the same company or by different companies. It is easy to understand what are the difficulties of controlling the flow of this sequence of operations in terms of both production and logistics.

All these steps constitute what is called a **process**; the techniques implemented fall under the science of **chemical engineering**. Each step, called **unit operation**, is a **technological building block**: we can mention distillation, crystallization, agitation, liquid-liquid extraction, etc. Chemical engineering and unit operations are the subject of Chapter 4 of the first volume.

Chemical engineering is the art of efficiently transforming raw materials into commercial products (Richard Darton in (Dal Pont 2012)). The process industries, the enormous variety of which has been highlighted, implement a multitude of processes with their own characteristics.

3.1. Anatomy of a process

The following examples describe the main characteristics of the processes cited in this book:

– Chapter 1 (Volume 2) focuses on biochemical processes with specific characteristics since they implement living organisms. This is a rapidly evolving industry that follows the progress of genetic knowledge;

– Chapter 2 (Volume 2) addresses the process which consists of making steam react to natural gas (NG). The multiplicity of the possible processes should be pointed out: this is typical of the process industries. A product can be achieved from different raw materials under different reaction conditions, including temperature and pressure. The process of making hydrogen involves different stages: NG hydrodesulphurization, pre-reforming, reforming (which is at the core of the process), separation of hydrogen from the CO_2 by-product and purification. The reforming furnace is a complex piece of equipment that can have several configurations. It is modeled on computers to improve its operation and design. The catalyst, the purpose of which is to accelerate the reaction rate, is obtained by extrusion or pelleting. Finally, it should be noted that reaction conditions can be modulated according to the customer's needs, by changing the steam-to-carbon ratio or the reaction temperature. In general, the complexity of the process and its high technicality is apparent. The tubes of the reforming reactor are a critical piece of equipment due to their cost;

– Box 2.6 (Volume 1) describes Vegan® technology for the production of bio jets and diesel fuels from lipids contained in vegetable oils, animal fats, algae, etc. The transformation is achieved in two stages; hydro-processing (hydrogenation), followed by hydro-isomerization. The technicality of the trickle bed-type reactor and especially the complexity of the gas/liquid distribution system at the reactor inlet is a striking feature.

3.2. Process evaluation

It is expected that a process be able to produce a profitable product, which generates a profit after it has been possible to recover in a reasonable time (return on investment time) the sums committed to research and to the investment in the production tool, while observing the criteria for sustainable development. **The quality of the process implemented is the *prerequisite* for its profitability**.

The raw materials involved, the energy consumed, the equipment (chemicals cannot be produced without equipment!), the manpower required for the operation of production tools – which are essential components of the total cost – depend on the process chosen.

The profitability of a product depends on its selling price, on the amount of the investment and operating costs, including distribution. There are often considerable differences in the amount of investments (CAPEX – capital expenditures) for the same process and same tonnage; the quality of the engineering company chosen for the implementation of a project may be the reason behind these differences. Operating costs (OPEX) can also vary significantly; we will revisit this topic in Chapter 4.

In recent decades, the concept of **sustainable development** has become a major topic of interest in the industry. Industrial disasters such as those of Seveso (1976), Bhopal (1984) and AZF (2001) all contributed to this. The concern of citizens in developed countries about the supposed or real impact of chemicals on the environment and on their health does not stop increasing. The concern is very strong regarding pesticides (see Chapter 2 of Volume 1).

Paul Anastas developed the concept of *Green Chemistry* in his book in 1998 and *Green Engineering* in his 2003 book (Dal Pont and Azzaro-Pantel 2014). The FFC (French Federation for Chemical Sciences) launched a program in 2008, "Chemistry for Sustainable Development", which did not survive it. Professor Jean-Claude Charpentier has highlighted in many articles and particularly in (Poux 2010) the fact that the commercial product and the process are intimately linked and that companies strive for *sustainable products*. Box 2.2 (Volume 1) by Catherine Azzaro-Pantel points out that the assessment on the environmental impact of a product during its lifecycle (LCA) *also* includes the impact of its production system.

The environmental assessment of a process is therefore part of this systemic approach: production, distribution and consumption.

The role of the process engineer is, starting from laboratory results, to define the industrial tool in order to acknowledge as soon as possible, with the highest accuracy possible, the costs of the investment and the product cost price. This is what the industrialization process is all about.

3.3. Industrialization process

Industrialization can be defined as the means and methods necessary for shifting from research and studies to an industrial tool (workshop-factory-plant) capable of delivering a marketable product meeting a number of defined specifications, in accordance with the budget, deadline and company ethics (Dal Pont 2012). This is an important act of industrial life: it is the culmination and realization of research and reflection works devised by the strategic plan.

Industrialization consists of two phases of a different nature:

– a study phase, known as the development phase, of a highly chemical nature, which consists of acquiring bases for industrialization;

– an engineering phase that is an implementation phase whose objective is the construction of the industrial tool. Its quality depends heavily on the previous stage.

3.3.1. *The foundations of industrialization*

Jean-François Joly describes this phase in detail in (Dal Pont 2012). We will give a brief summary of this here.

The aim of this phase is to establish a preliminary process diagram with a material balance and an energy balance (enthalpy balance), a description of the main material and especially of the critical material (often the reactor, as we saw in the previous chapter for hydrogen production).

This phase must confirm the **technical economic** and **environmental** feasibility of the process, and thus its viability. The cost price of the product and the amount of investment are assessed, and the raw materials are defined. Technological challenges are detected and unleashed. The **robustness** of the system is confirmed, which is a fundamental element. For example, the lifespan of the catalyst is determined and its cost reduced to an acceptable value, while the accumulation of impurities is controlled. Building materials are wisely chosen to withstand corrosion and abrasion. Extrapolation criteria are defined. By-products and waste products are not omitted and are an integral part of the process.

The process diagram is the cornerstone of this basic step, strictly speaking. It will be refined with the acquisition of knowledge. The latter is obtained from pilot tests, in cold mock-ups. The kinetics of the reaction, the "intrinsic kinetics" to quote Jean-François Joly, are determined from microreactors. The modeling of the reaction system provides a means to choose the most suitable reactor technology.

Modern simulation tools are used to optimize the process in order to determine recycling loops. **MSO** (modeling, simulation, optimization) implements commercial or proprietary software (Poux 2010). However, pilot costs are increasingly high; the current trend is to **miniaturize** testing equipment and **robotize them**. Catalysts are being increasingly tested in microreactors.

The culmination of this phase is the **basic file of industrialization**. This file will constitute the starting point for the engineering phase. It is assumed that the file is complete enough so that nothing unexpected happens at the start of the installation.

Other activities can take place in parallel to this essentially technical phase:

– commercial validation, in terms of prices and sales volumes, evaluation of the competition (benchmarking). BCG or SWOT techniques can be used (Dal Pont 2012);

– protection of the intellectual property using the Soleau envelope which testifies, at minimal cost, to prior records of the invention and identifies the author, or by filing a patent, which is a costly operation;

– validation of the product in terms of use value to potential customers who will *sample* it;

– regarding pharmaceutical drugs, clinical trials are to be conducted which, in addition to verifying the activity of the product, are intended to determine its safety (side effects): this is an essential precondition for marketing authorization (MA). This authorization is issued in the United States by the FDA (Food and Drug Administration) and in France by the ANSM (*Agence nationale de sécurité du médicament et des produits de santé*) upon presentation of the file. The complexity and the cost of the files is assessed; moreover, the MA can be suspended in the event of adverse side effects occurring after a few years of marketing;

– pursuing partnerships to share investment risks and to make further developments.

3.3.2. *Realization (project) engineering*

This comprises all the activities that aim to build a project: a plant, a bridge, a building. It transforms financial resources, **investment**, into a physical good.

A **owner** company rarely has all the necessary competencies with regards to skill or volume at any given moment to achieve an investment. This is directed to a **contractor** under the name of an engineering company.

These engineering companies offer their customers:

– know-how in a variety of industrial fields (oil, chemistry, pharmacy, etc.). They know the techniques;

– human resources in about 20 different trades: process engineers, project engineers. Specialists in civil engineering, electricity piping, rotating machines, insulation, boiler making instrumentation to name but a few. Not forgetting the construction managers, cost-estimators and so on (Charrier 2001).

A large number of companies provide **EPC** (Engineering Procurement Construction) services:

– *engineering* involves the **study** in line with what has been defined for the bases of industrialization. This is the field of the **process engineer** who is at the core of the industrialization process; he will define the PFD (Process Flow Diagram), P&IDs (Process & Instrumentation Diagrams) and the specifications (SPECS) of the main equipment. The **process engineer** will initiate the project. Engineering also involves detailed engineering (see below);

– *procurement* consists of the supply of equipment (distillation columns, exchangers, etc.), tendering, construction supervising, bringing equipment to the site and verifying the performance once in place;

– *construction* corresponds to the activities at the future plant site. Those responsible must have strong nerves to deal with the daily problems which inevitably appear, whether with subcontractors, workers or the project owner, and must always be critical.

Engineering companies work on a **project basis**.

3.4. The concept of the industrial project

The industrial project is a unique specific activity that has a beginning and an end, and is therefore limited in time. It implements human and financial resources, multidisciplinary teams and equipment for a limited period of time.

The players involved in the project have to bounce back when the project comes to an end, clearly often they have to find another project. The project creates change and affects the environment in the broadest sense.

To simplify, we will take the construction of a house as an example of a project to define the basic notions:

– the **owner** sets the objectives, the budget envelope and the desired deadlines;

– the **contractor** designs the work with the owner (the client) and realizes it;

– a project is above all a **team**, under the authority of a **project supervisor** who coordinates all of the phases and makes the necessary decisions in order to run with no problems;

– a project consists of a number of **tasks**: buying land, foundations, walls, framing, roof, electricity, plumbing, painting, etc.;

– the **tasks** are carried out in a logical order according to **planning** (usually the installation of the roof requires the walls to be mounted);

– the **functional specifications document (FSD)** is the written document that reflects the client's expectations in terms of expected functions and services. It is very often called the project's scope: it mentions the constraints (examples: budget envelope, security aspects). It is jointly defined by the project owner and the prime contractor. Its aim is to avoid **misunderstandings** and thus to define the scope of the project: number of rooms, global surface, number of floors, etc.;

– a **contract** binds the owner to the prime contractor: this is a key element of the project. We can mention a turnkey contract (also called a lump sum contract) or even *cost + fee* (cost plus fee contract). The best contract is the one that is a win-win for the owner and the prime contractor, a contract that forces them to cooperate effectively to reduce costs and delays and share the resulting gains;

– the client expects that the work handed over to him/her not exceed the budget, and be delivered to him/her on the scheduled date and in accordance with the FSD. This is not always the case!

– the terms **costs, schedule, quality** indeed define what a project is: **there is no project without a boss, and without allocated resources**.

3.5. Typical organization of an industrial project

Figure 3.1 represents the typical organization of a project. The project owner, the prime contractor and the operator are the key entities of the project.

It is important to determine at which point the engineering company is involved in the development phase: if it intervenes too late, it will facing a "*fait accompli*" and complain that it has not provided any know-how to its client; if it comes too soon, the client will incur costs that it will not always deem justified. **An engineering company essentially charges for its time**.

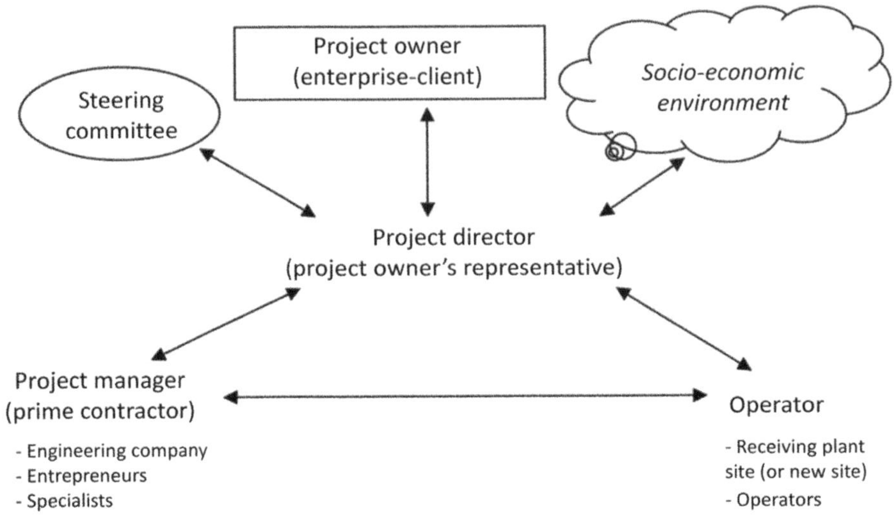

Figure 3.1. *Typical project organization*

The same is true of the future operator: it must be associated with the project. It will be the one inheriting it! The future operator may be accompanied by the maintenance engineer, the instrumentation manager and other key people of the plant. Those who do not participate in the project in a timely manner and who inherit it will not hesitate to make often scathing criticisms and ask for modifications.

The steering committee consists of **stakeholders**. This relatively recent term represents entities that have a **stake** in the project or that are affected thereby. Without being restrictive, and bearing in mind that it all depends on the situation, let us quote:

– the owner's representative, who is generally responsible to its shareholders. He/she must have a position of importance in the hierarchy to be able to implement the appropriate measures;

– project partners (investors, JV representatives);

– representatives of the company's key functions as well as of its needs (HRD, industrial management, business owners, etc.);

– the administration that has a say in building permits and in various regulations. It can play the role of facilitator with central agencies;

– representatives of associations;

– lucky residents that benefit from the jobs created by the investment but who fear the disruption brought by it.

Finally, it should be noted that a project is still part of a socio-economic environment that can be volatile. The **scope** of the project (a term that is becoming increasingly important) can be affected: the project can then return to a zone of turbulence. The **change orders** resulting there from are very expensive. Engineering companies know how to get paid!

3.6. The stages of an industrial project from the engineering perspective – validations

Engineering companies, similarly to every organization, implement their own working methods, a specific business jargon that, with the help of globalization, has become international: thereby, the terms *basic*, *detailed*, etc. are heard all over the world. This terminology should be known.

It is important that the project owner, although they have power conveyed by money, understands how their prime contractor works, how they are organized; the article (Charrier 2001) can be of great help with this.

An owner-manager ought to know as soon as possible the costs incurred in the installation of the project they are managing, the time it will take to build and start it. They will want to know if the business is profitable, which is an internal business matter within the company. They are accountable to the business manager, to their hierarchy. They appeal to a prime contractor (engineering company) for the reasons mentioned above. However, studies are expensive and take time.

Table 3.1 reports the steps that (it is given as an estimation for the large chemical industry):

– order of magnitude;

– feasibility;

– preliminary project;

– basics.

Then, after the basics, follow the detailed studies that involve the different trades and the construction.

At each stage are associated diagrams and plans that will become increasingly more accurate and more numerous, depending on the progress of the studies. The accuracy of the PFC (probable final cost of the Investment) will depend on the stage, reflecting the degree of progress. For example, Figure 3.2 shows the evolution of the PFD to the PID (piping and instrumentation diagram) of a simple heat exchanger.

Nature	Accuracy[1]	Cost % PFC[2]	Objective	Necessary documents	Comments
Order of magnitude	± 50 %	Variable	Orientation/Selection Continuation/Discontinuation of R&D	Simple process description Preliminary diagram Material approached	"Back-of-the-envelope" estimate
Feasibility	– 20 % + 40 %	Variable	Verify whether the process is viable	Process flow diagrams Equipment list	Allows detecting what remains to be done, thus guiding the "process development" phase
Preliminary project	± 20 %	0.5 to 1%	Verify (economically) whether the basic study is justified Budget registration	As above + 1st edition of the P&IDs[3] Site layout Specifications of the main equipment	The site is defined in terms of the project "philosophy". Building, instrumentation, electricity
Basic study	± 10 %	3% to 7%	Request Credit authorization	As above + "frozen" P&IDs Defined equipment	This dossier accompanies market studies, risk analyses, audits
INVESTMENT APPROVED – FUNDING GRANTED					
Detailed Engineering → Construction → Tests (reception) → Start-up → Industrial operation 12 to 15% of the PFC					

(1) The estimated cost is generally given within a high and low range. Therefore, a cost of 100 (for example, 100 million euros) within a range of – 20%, + 40% means that it is estimated that the PFC of the investment will be between 80 and 140 million euros. The range depends on the company policy A number of companies choose – 10%, + 20%.

(2) PFC = probable final cost.

(3) P&IDs = pipe and instrumentation diagrams (all pipes, valves, dimensions, nature, etc.) + detailed instrumentation.

Table 3.1. *The stages of an industrial project from an engineering perspective*

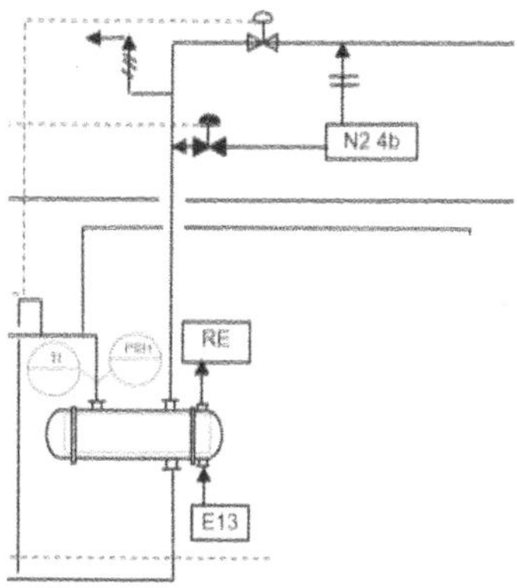

PFD excerpt

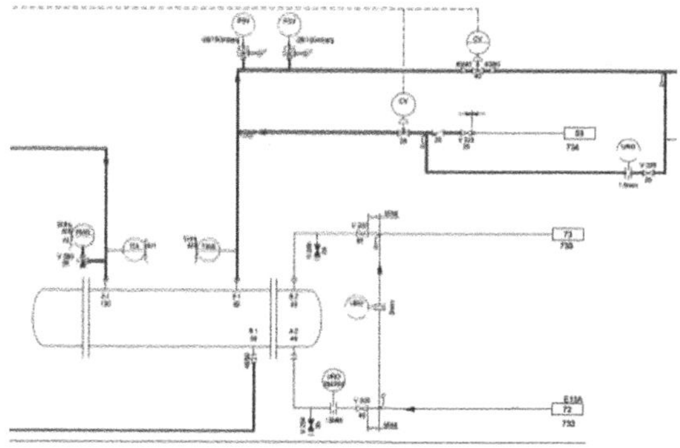

PID excerpt

Mock-up excerpt

Figure 3.2. *From the process diagram to the mock-up (document provided by ERAS). For a color version of this figure, see www.iste.co.uk/dalpont/process2.zip*

The *basic engineering* is the file supporting the loan or funding application, which is a fundamental step, if ever there was one. A greater precision would require further studies, more invitations to tender, it will be more expensive and will take much longer.

Table 3.1 is a good indication and estimation that a basic study costs 3 to 7% of the PFC; the basics of a workshop of 100 million euros can be estimated between 3 and 7 million, which is a significant sum; hence the importance of the steps that precede.

Validation is a key process in the industrialization process. Responsible industrial management requires that project validation be done as the project, initially unclear most of the time, takes shape both from a technical and commercial point of view. The product sale price and expected volumes are part of the project. Serious validation answers three questions: *should we proceed according to the initial scope? Should the content (scope) of the project be modified? Should we stop it?* This technical and economic validation leads to a decision that is always difficult, sometimes dramatic. Stopping is a courageous operation but one that affects the teams that have been engaged and writes off all the money already spent. On the other hand, time is an impartial judge: from the start-up phase, we shall see if the assumptions and the technical choices carried out have been made wisely.

3.7. The tools of engineering project management – related activities

Some of the techniques implemented by engineering companies, excluding conventional techniques used by the process engineer and those of the various trades, are listed below as an indication.

3.7.1. *Process conceptualization: making it visible*

It is essential that the company's stakeholders have a vision of the project, to which it commits in terms of cost, risk, competitive advantage, impact on the company in general and on its business in particular. In addition to a simplified process diagram, a block diagram (Dal Pont 2012) made using the WBS (Work Breakdown Structure) technique, is properly suited to this visualization exercise.

Figure 3.3 from (Dal Pont 2012) gives the WBS of a latex workshop. A quick glance allows us to understand that the dangerous nature of the workshop lies in the butadiene part assimilated to a flammable gas, and its technical nature in the design of the polymerization reactor; in addition, the supply by ship also raises particular concerns about the logistics.

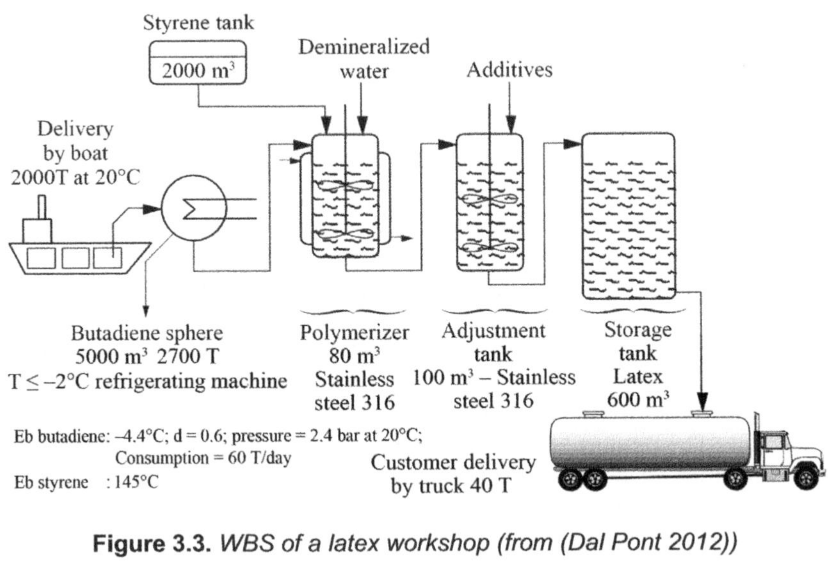

Figure 3.3. *WBS of a latex workshop (from (Dal Pont 2012))*

3.7.2. *Project management*

Project management is clearly the core business of engineering companies. This concept would require a very long development. So very briefly, it includes setting up teams with the skills required by the project, activating them, defining tasks, planning, supervising and monitoring the activities, as well as controlling costs and quality.

3.7.3. *Reporting – executive summary*

Reporting is an essential part of project management, we set it apart given its importance. It is the work of both the prime contractor and the project owner.

In general, on a monthly basis, it reports the activity of the past month, includes expenditure statements and the degree of progress, assesses what has worked well, the difficulties, the problem areas that were found and the solutions implemented or to be implemented. It provides an overview of the activities of the following month, the financial commitments to come. The project owner's reporting is based on that of the prime contractor, but it provides information for the company's executives, such as the presidency or the Board for major projects.

The **executive summary** is therefore the succinct document that allows managers to follow the project. In the event of a cost slippage, it would be in everyone's interest that the CFO (Chief Financial Officer) be alerted in due course. **Uninformed people quickly become detractors**.

3.7.4. *Other concepts*

– Value analysis, functional analysis.

– CAPEX-OPEX evaluation and optimization.

– Site selection.

– Outsourcing.

– Obtaining permits (construction).

The essential qualities of a project manager are leadership, recognized competence and unwavering resilience, including health.

Figure 7.4 is a summary of the stages of a project and the validation processes; the surface of the "bells" gives an idea of the resources involved. The initial stage of a project can be carried out by a few people at the benchtop, while thousands of workers can be active at a construction site.

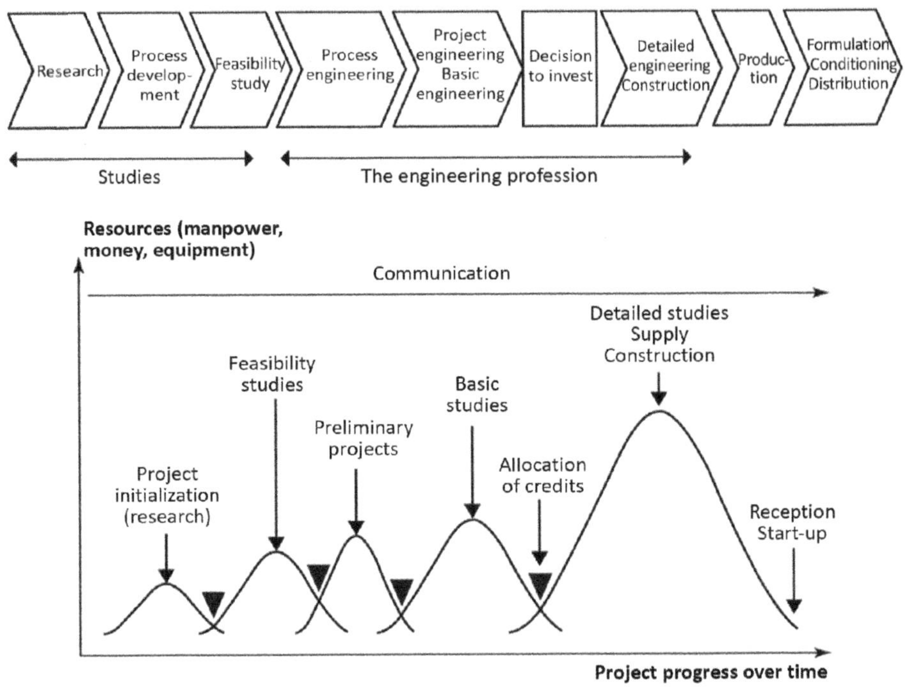

Figure 3.4. *The stages of a project. Validations (black triangles): proceed, proceed by modifying, stop*

3.8. Process intensification (PI) – miniaturization

Process intensification is the subject of Box 5.1 in Volume 1. This promising concept was initially confused with process miniaturization, that is the use of micro-channels or micro-structures to take advantage of the surface/volume effect. Surfaces increase by squaring the size and volumes by taking the cube of the size; this simple notion is sometimes at the root of many setbacks. An agitated reactor of one cubic meter that can be cooled by its double shell can no longer be cooled if it is 10 m^3-internal coils or an external exchanger must be added. Chapter 2 concludes with the utilization of micro-structured reactors using additive manufacturing (AM) – see Figures 2.27 and 2.28.

The article by Christophe Gourdon (Gourdon 2016) is a good review of PI.

Very succinctly, regarding **reaction**, it is a matter of finding the limiting effects (heat transfer, material transfer, chemical kinetics) and mitigating them, by activation of the reaction through an appropriate catalysis, by ultrasound or using new types of reactors (rotating disc reactors), etc.

Functional analysis is carried out on equipment pieces, which are considered to be **technological building blocks**, and efforts are made to combine their functions. A typical example is the reaction-distillation combination; the reactor is included in the separation column, the normal function of which is the separation of products generated by the reaction.

A major objective of the pharmaceutical industry, which most often operates discontinuously (in batches) is to shift to continuous operation, and if possible, into microstructures if the fluids do not contain solid particles that would clog them.

3.9. Investment/sales coupling – modular construction

Investing is a risk! Among the usual risks associated with projects of any kind (Poux 2010) we can identify technical risks: start-up difficulties, failure to achieve performance (efficiency, product quality). There are also commercial risks, often major risks. Simply put, sales don't always happen; they usually take a few years before levelling off, to decrease at the end of the product lifecycle.

The question for the industrialist is whether to invest "up-front" in a facility of the capacity necessary to satisfy sales at the maximum volume expected, and therefore to invest "massively", or to try to invest in a smaller capacity with a reduced investment ... and see what happens next. The ideal for industrialists is to *align* capital expenditures and sales, and therefore revenues, to preserve their cash flow.

Several solutions are possible. Manufacturing can be outsourced to a third party that is specialized in it, in other words a third party that possesses adequate multi-purpose equipment. Another solution consists in investing in a small workshop and in adding a twin workshop when the first unit becomes saturated, and so on.

Modular construction, for its part, consists of prefabricating "modules", namely sections of the installation: one module might hold the reactor, another a distillation column, etc. The benefits of modular construction are widely highlighted in Box 3.1. Some companies offer functional contained modules, containers on skids: reactors, crystallizers, filters, wastewater treatment, mini chlorine or phosgene production facilities, etc. The construction then becomes similar to a Lego game, the modules can in principle be reused.

3.10. Circular industrial economy – platforms – centralization – decentralization

The selection of a production site is often a delicate operation because it is irreversible (Dal Pont 2012). Very large companies in Germany, the United States, China, etc. sometimes have sites with tens of thousands of employees. The new workshop usually finds everything it needs in terms of basic products, utilities, logistics and skills.

These platforms have undeniable advantages: the proximity of the workshops saves a significant number of transports (thousands of trucks a day, according to a German company) and the by-products of workshops become the materials for nearby workshops. The same can also be said of utilities. Medium or low-pressure by-product steam from a high-pressure user easily finds its use in users who do not require high temperatures.

Box 3.2 describes the Pont-de-Claix platform, which implements extremely dangerous products, including phosgene. The latter has in the end become its major asset, although the platform is located very close to the city center.

In recent years, the issue of centralization and decentralization of production resources has begun to arise. In the spirit of sustainable development and in order to put human beings at the center of the device, to follow the formula already stipulated (see Chapter 6), the issue concerns industrial entities based on natural resources of human size. Some may say right in the city because we now know how to build safely.

3.11. Overseas operations – technology transfer

Globalization and the incredible rise of Asia-Pacific countries, including China, Singapore, South Korea, Hong Kong, Taiwan, India, Indonesia, etc., have led developed countries (including Japan) to establish industrial facilities at these locations, which is a prerequisite for doing business there. The first phase, which is a matter of trade (export), must quickly contribute to a local production required by the country under consideration to create jobs and integrate into the value chain. In addition to these political considerations, operations abroad are necessary to reduce the length of the supply chain, to adapt the product to local needs, as well as to facilitate after-sales service and knowledge of the field. We developed these concepts in (Dal Pont and Azzaro-Pantel 2014).

The exported technology needs to be reviewed in most cases. Examples include:

– the design conditions (from – 40°C to + 40°C in the Urals);

– the specifications of the commercial product, *slightly* different raw materials;

– the respect for local culture (work organization);

– regulations.

Figure 3.5 attempts to show that the transferred technology, represented by the rectangular grid, is part of a larger project that takes local constraints into account. As a result, the project manager holding the know-how must associate with a local project manager and an external affairs manager to ensure contacts with the local government and stakeholders.

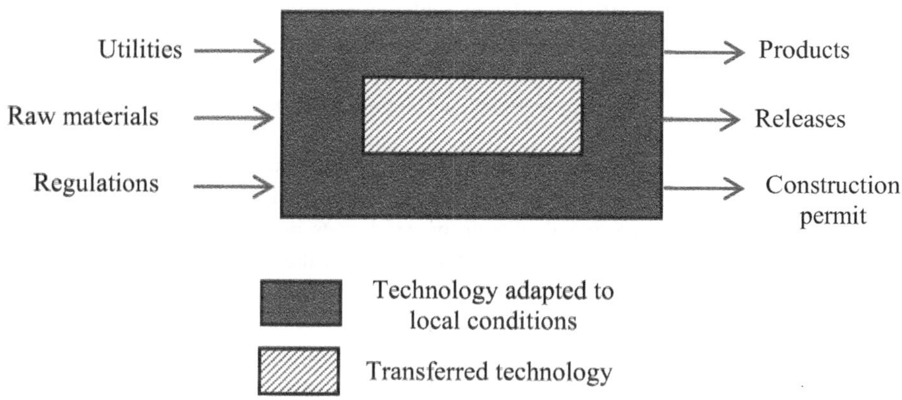

Figure 3.5. *Technology transfer*

3.12. Conclusion

The engineering profession is a difficult, sometimes unrecognized profession. A systemic vision of the project, a global vision and a focus on every detail are equally necessary!

The relationship between the project owner and the prime contractor is often conflicting because of the lack of knowledge about the working methods of each part. Should the prime contractor (engineering company) be limited to a role of assembler, in other words should they merely juxtapose pieces of equipment, or should they cooperate with the owner to find innovative solutions? This is a difficult issue that involves risk sharing. We hope that the above will help to find solutions that also include equipment manufacturers.

3.13. Boxes

Modular installations

Advantages over a conventional stick-built facility

Modular manufacturing offers an alternative to construction by using its specialized methodology – modularization, prefabrication and pre-assembly – to positively impact safety, quality, schedule and cost.

This system provides an efficient solution, with low operating costs, countless advantages due to its ease of assembly and use, as well as low maintenance and a small footprint.

Two factors drive the success of a project: a design-build approach and early involvement.

For turnkey engineering, modular manufacturing and construction service providers, it is necessary to work hard from the start of the projects to plan existing site conditions, necessary adjustments to the workforce or volatility in material costs during project execution. With detailed planning, experience and early communication in the project planning phase, it is possible to offer clients a total installed cost.

The five-phase project methodology divides a project into manageable segments to better guide the project planning process up to the final functional hand-off to clients:

1) preliminary engineering;

2) detailed design and engineering;

3) procurement and prefabrication;

4) installation;

5) start-up and commissioning.

Off-site construction – including prefabrication, modularization, pre-assembly or off-site multi-skills manufacturing – is becoming a critical method for delivering projects faster, safer and cheaper than in the current engineering and construction environment where labor-related constraints are increasingly significant.

Figure 3.6. *Modules being transported to site. For a color version of this figure, see www.iste.co.uk/dalpont/process2.zip*

Figure 3.7. *Roeslein worked directly with the refinery and EPC firm on the detailed engineering and the manufacturing of 28 process modules and 11 pipe-rack modules for this Midwest refinery. The project consisted of 4,500 linear meters of pipes and 600 tons of steel. For a color version of this figure, see www.iste.co.uk/dalpont/process2.zip*

Over time, it is estimated that engineering and construction industries can move from off-site manufactured construction to true production-oriented manufacturing. However, industry as a whole is still stuck in its old ways and will have to overcome some significant obstacles to truly evolve and transform itself in the years to come.

The drivers of modular construction are more complex than the simple robustness of their structure and must be taken into account when assessing the advantages and disadvantages of modularization.

For example, in order to convert 112,000 tons of pig manure per year into pipeline-quality renewable natural gas (RNG), turnkey modular design provided the design and manufacture of a flare, compressor and membrane skids as well as the offsite design and manufacturing of all electrical, instrumentation and programming controls required for each of these sections.

Due to the remote location of the project, a modular design-build approach has allowed the production to take place at a 37,000 square meter site in Red Bud, Illinois. The project will produce up to 2.2 billion cubic feet of pipeline-grade NGV, equivalent to 64 million liters of diesel fuel per year.

The modular construction of this alternative natural fuel facility allows for it to be economically replicated worldwide at a lower cost, faster start-up and simplified operations, and offers alternative solutions tailored to the client's needs.

The implementation of prefabrication and the initial design of the modularization played a major role in the overall project execution strategy and helped reduce costly work hours on site while increasing productivity.

Quality: in-field manufacturing quality will often be lower than the modularization quality, which is superior to it because of the quality controls that can be carried out frequently and efficiently in the workshop before installation.

Safety: modularization reduces the number of hours per person on site, reduces coaching hours, work at heights and limits the number of different trades. The risk of an accident is significantly minimized due to the reduction in the number of workers and on-site operations.

Planning: the time for the installation and connection of the modules is significantly less than that of on-site construction. All overhead costs, such as those of tools, personnel management, are also reduced.

Costs and inefficiencies: on-site projects often require work at 6 to 9 meters high above ground level. This is a major safety barrier for on-site work, including all equipment, tools and procedures needed to meet OSHA requirements.

Weather: the weather is unpredictable in all seasons and can be a cause of danger. Weather conditions may affect productivity. Bad weather may lead to increased costs, increased hours, reduced quality and reduced safety. At Roeslein, every module is built indoors, at 21°C, without rain, snow, mud or frost, all year round.

Field-Change Orders: escalating costs and the risk of overrunning costs for contractors are fairly common in stick-built projects. The cost of unforeseen events must be incorporated into on-site projects. The costs of modular systems are fixed from the outset and there is no modification order unless the scope of the project is altered.

Construction Cost Comparison Modularization with ROESLEIN *versus* on-site construction: modularization is often compared to traditional "stick-built" construction systems solely on the basis of labor and equipment. A more accurate comparison of the total cost installed should include the differences in indirect costs associated with on-site labor, timing, quality, productivity, safety and general conditions.

The disadvantages of traditional stick-build construction compared to modular construction are presented hereafter:

– on-site storage/field/circulation area;

– control of equipment when it arrives at the site;

– lack of staff at times;

– scaffolding, cranes and pods costs are often underestimated;

– need for offices and site trailer space, for office supplies necessary throughout the project;

– installation and dismantling of construction huts for welders and for housing equipment;

– site security and fencing costs are reduced when the installation time of modular systems is faster;

– incoming freight for the shipping of every piece of steel, pipe, nut, bolt and tool to the site;

– non-destructive tests (hydraulics, X-rays, etc.) can be expensive;

– procurement, shipping and material management costs in stickbuilding are not always included in the overall estimate.

Conclusion

Modular plants represent a solution between batch and continuous processes. They are able to work continuously and flexibly.

Additional photos

https://www.dropbox.com/sh/dad5ql3vjoj4kzx/AADTHixxUhQTYtStW4h7ewZNa?dl=0

Roeslein & Associates

Roeslein & Associates is a company founded in 1990 by Rudi Roeslein and Fritz Dickmann. Located in St. Louis, Missouri, it comprises 600 professionals from around the world. It has engineering offices in Denver, Colorado and in China.

John DeMoulin

Vice President, Head of Business Development and Marketing at Roeslein & Associates, John DeMoulin has nearly 30 years of combined experience in Business Development and Marketing.

At Roeslein Associates, he is responsible for client project implementation, including initial business development, proposal generation, contract negotiation and active financial management. He participates in developing new market opportunities. He works directly with the ownership group in the development and implementation of the corporate business plan, sales targeting and forecasting for new account/market penetration. John also spearheads the development of corporate marketing literature, including newsletters, direct mailings, company brochures and website development.

jdemoulin@roeslein.com

Box 3.1. *Modular installations (John DeMoulin)*

Circular economy: the Vencorex platform in Pont-de-Claix

The Pont-de-Claix plant is a century-old plant created in 1916 to manufacture chlorine along with the co-production of soda and hydrogen by electrolysis of brine and white coal (Hydro-electricity), then in its infancy, for military purposes. It has undergone major transformations over its long history, but it has been able to maintain a master plan, that is to say, an exemplary logistical organization.

Figure 3.8 illustrates the main interactions between on-site entities and Table 3.2 lists their main product exchanges.

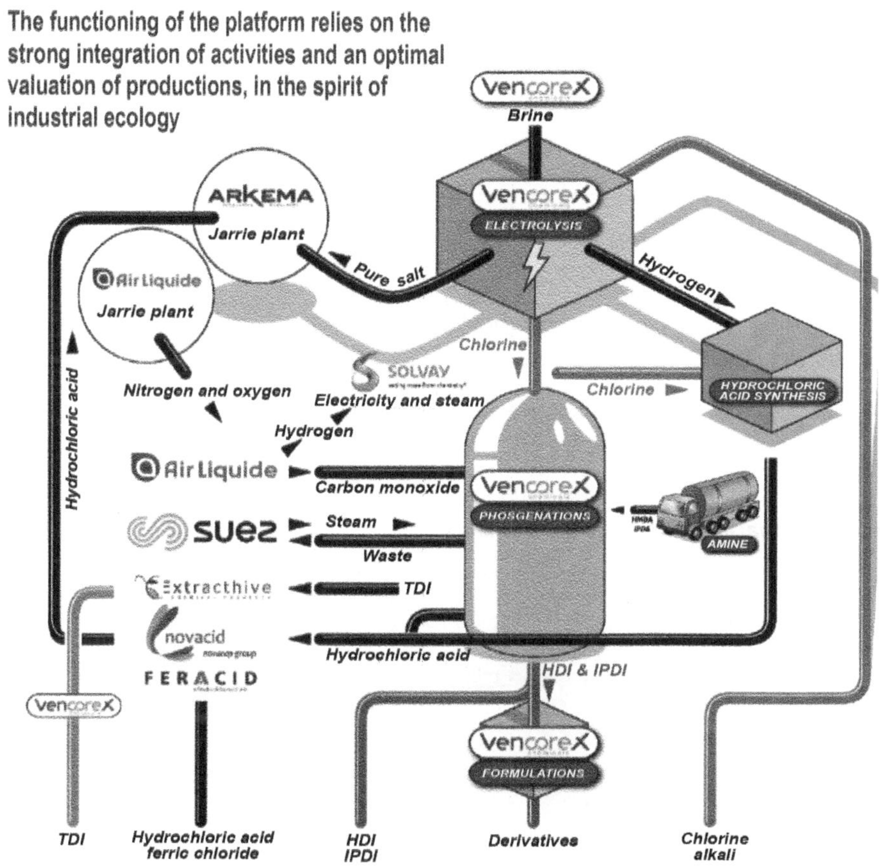

Figure 3.8. *Interactions between site production units. For a color version of this figure, see www.iste.co.uk/dalpont/process2.zip*

Production unit	Process	Basic products (raw materials)	Co-product 1	Co-product 2	Co-product 3
Vencorex electrolysis	Membrane electrolysis	Brine + electricity	Chlorine	Soda-ash	Hydrogen
CO/H_2 Air Liquide	Natural gas oxidation	Methane + oxygen	Carbon monoxide	Hydrogen	–
Vencorex HDI 1 and HDI 2 ("bubbles")	Phosgene production	Chlorine + carbon monoxide	Phosgene (for HDI and IPDI isocyanates)	–	–
Vencorex HDI 1 and HDI 2 ("bubbles")	Pressure phosgenation	Phosgene + amines	HDI or IPDI	Hydrochloric acid gas	–
Hydrochloric acid synthesis (after electrolysis)	HCI burners	Chlorine + hydrogen	HCI solution (hydrogen chloride)	Steam	–
CEV Solvay	Cogeneration	Methane, demineralized water	Electricity	Steam	–

Table 3.2. *Key processes and products implemented in Pont-de-Claix, interactions*

Essential characteristics of the plant

Figure 3.9. *In the foreground, the electrolysis room; in the background, a phosgenation "bubble". For a color version of this figure, see www.iste.co.uk/dalpont/process2.zip*

Initially in the countryside in the early days, the plant is now located in an urban area: the workers needed to be accommodated and the expansion of the city of Grenoble had to be dealt with. It is also situated in a seismic zone. It manufactures and utilizes dangerous products as shown in the previous table.

Safety and energy balances are both constraints and opportunities for the platform, which has been able to preserve and develop a first-rate know-how, the latter has been a prerequisite for proceeding with its operations.

Safety

In the chemical industry, the risks are mainly related to the dangers of the products when handled and the processes implemented. The safety of people and facilities is a top priority, and prevention is the first priority. The aim is to control these risks, in order to avoid accidents. Preventive measures are preferred over the human, organizational, material and technical aspects.

Any emanation of toxic fumes must be contained within the plant and not affect its neighbors.

Process safety, risk reduction at the source

The control of technological risks is based on prevention and risk reduction at the source. Each production facility is subject to an operating authorization issued by the prefectural authority. As early as project design, the choices are orientated towards the best available technologies, presenting notably every possible guarantee in terms of reliability, safety and respect for the environment. Hazard studies form the basis of the risk management approach. These studies are reviewed and updated regularly, under the supervision of the DREAL (French Regional Directorate of Environment, Planning and Housing).

Safe operating conditions

Manufacturers define the operating conditions that allow the safe operation and the use of manufacturing facilities. In addition to process-related aspects, three areas are examined: the equipment (equipment robustness, industrial IT – information technology, compliance with standards, etc.), human resources (skills, staff training, empowerment, etc.) and the organization (procedures, instructions, control, inspections, etc.). Safety barriers are also determined to avoid drifts and limit the consequences of a possible incident. Safety controllers in facilities can be “sorting redundant” to ensure their operation, even in the event of a failure on one of the circuits. They are systematically separated from the steering systems in units for maximum reliability and efficiency. For each production workshop, a hazard study is carried out. The platform companies comply with these orders and review hazard studies according to time schedules set by the administration.

Process safety: an example of the containment enclosure

At the beginning of 1987, the construction of the new phosgenation unit was launched. It produces HDI (hexamethylene diamine), an isocyanate monomer destined for the market of high-end polyurethane paints and varnishes.

The choice of the process, phosgenation under pressure, drives the plant into adopting new provisions in terms of process safety: deviation studies, instrument duplication, earthquake resistance, etc., and of course, the famous double containment enclosure colloquially referred to as the "bubble". This nuclear-inspired safety device is designed to contain possible toxic gas leaks inside a watertight building.

The "bubble" provides maximum protection for employees and neighboring populations. However, working conditions are demanding for staff: in order to work safely inside the containment enclosure, the operators must equip themselves with a self-contained breathing apparatus (SCBA). This material offers protection in the event of a toxic gas leak. A specific three-day training course is provided to the staff concerned.

Staff safety and health

Each staff member is trained specifically to the workstation that he/she is assigned to hold. Several competency and skill levels exist depending on the occupational activity (job) and the nature of the risks. The entire process of certifying an operator can take several months or even more than a year when there are multiple tasks.

In order to continuously progress in terms of safety, awareness-raising efforts, particularly focused on behavior (collective and individual), are undertaken in order to achieve the objective that must be shared by everyone: "zero accidents."

Regular retraining and safety exercises are carried out to train and maintain knowledge. This approach also applies to the staff of external companies who collaborate in the workshops.

The subcontracting companies on the platform are associated with actions aiming to improve working conditions, safety and environmental protection, in accordance with the regulations.

Companies involved in chemical facilities possess the MASE certification (French enterprise safety improvement manual), based on a system of continuous safety improvement. For staff, nominative clearances are required.

Internal controls (safety technicians, warnings, occupational safety committee, etc.) or external controls (DREAL, labor inspector, etc.) are intended to verify compliance with rules and procedures: instructions, working conditions, use of safety equipment, equipment compliance, etc.

In parallel with safety-related actions, the health of workers is subject to special medical supervision. The priority is to avoid exposure to products.

In case of emergency

The platform implements significant preventive measures to control industrial risk. In the case of consequences for humans and the environment outside the site, we speak of a major risk. Although the probability of an accident of this type is very low, specific provisions are in place provided by industry and government to address this type of situation.

Emergency plans

– The internal operation plan (IOP) of the platform can be operational 24 hours a day, 365 days a year. The IOP is the subject of a monthly simulation, with different scenarios (fires, gas leaks, transport of hazardous materials, etc.). Once a year, a larger scenario allows observers and external resources to be involved.

– The particular intervention plan (PIP) is the subject of civil safety exercises supervised by the prefecture, with the involvement of the various parties involved (industrialists, local authorities, law enforcement officers, public emergency services, etc.).

The means of intervention

– A company of volunteer firefighters trained in chemical hazards, 200 rescue workers, an integrated medical service, support teams (including one specialized in environmental matters).

– A computerized toxic gas alert system, tested weekly.

– A dozen emergency vehicles.

– A fire water system crisscrossing the entire platform.

– A "major risk" alert siren to the population, three "mobile warning equipment" vehicles, an automatic telephone alert system for the riverside town halls.

– A mutual assistance protocol with the neighboring Arkema plant at Jarrie.

Energy balances

The platform is both energy-producing and energy-intensive.

Phosgene

– Electricity and steam from Solvay's SPP (steam power plant) equipped with cogeneration turbines and steam boilers. The SPP supplies the plant and the EDF network supplies on-call electricity during peak periods.

– Steam originating from the heat released by the production of hydrochloric gas obtained by the reaction of chlorine with hydrogen.

Consumption

Distillations and evaporations are steam consumers as well as all conventional chemical machinery: pumps, agitators, compressors, etc.

Membrane electrolysis is highly electricity consuming.

Exchanges with metropolitan Grenoble

The platform is now connected with the networks of metropolitan Grenoble with which it exchanges heat in order to optimize and enhance the production and consumption of energy throughout the year.

The platform provides energy to the metropolitan area in the winter when heating needs are significant mainly during peak consumption. During the summer, the roles are reversed: the city provides energy to the platform, in the form of energy mainly coming from its urban waste incinerator.

The energy optimization of the site, which varies according to production volumes, climate, fuel prices, EDF demands, requires consummate know-how and resorts to sophisticated IT resources. This is almost daily management!

Advantages of the platform concept

In logistical terms, this industrial structure has the advantage of optimizing the flow of materials from one workshop to another, with direct links via pipelines, which limits storage and avoids transporting by road or rail certain hazardous materials with related GHG (greenhouse gas) gains.

It shares important resources in terms of safety and environmental protection in the broadest sense.

Its constraints become opportunities: we can talk about industrial ecology.

It is one of the few facilities to manufacture phosgene in Europe. Its culture of safety is the main guarantee of its sustainability: human beings are as usual at the center of the system!

Vencorex

Vencorex develops state-of-the-art technologies for the manufacture of aliphatic isocyanates, for the paint, varnish, coatings, high-performance adhesives markets used in automotive, aeronautics, electronics, industry, etc.

Patrick Pouchot

Patrick Pouchot is head of communication at the chemical platform of Pont-de-Claix.

Box 3.2. *Circular economy: the Vencorex platform in Pont-de-Claix (Patrick Pouchot)*

3.14. References

Charpentier, J.-C., Dal Pont, J.-P., Jullian, S. (2014). Industrialisation des procédés et usine du futur [Online]. *Techniques de l'Ingénieur*. Available at: https://www.techniques-ingenieur.fr/base-documentaire/procedes-chimie-bio-agro-th2/industrialisation-des-procedes-et-usine-du-futur-42602210/#presentation.

Charrier, M.-C. (2001). Réalisation de projets dans une société d'ingénierie [Online]. *Techniques de l'Ingénieur*. Available at: https://www.techniques-ingenieur.fr/base-documentaire/procedes-chimie-bio-agro-th2/industrialisation-des-procedes-et-usine-du-futur-42602210/realisation-de-projets-dans-une-societe-d-ingenierie-ag3300/.

Dal Pont, J.-P. (2012). *Process Engineering and Industrial Management*. ISTE Ltd, London and Wiley, New York.

Dal Pont, J.-P., Azzaro-Pantel, C. (2014). *New Approaches to the Process Industries*. ISTE Ltd, London and Wiley, New York.

Gourdon, C. (2016). Intensification des procédés – Fondamentaux et exemples d'industrialisation [Online]. *Techniques de l'Ingénieur*. Available at: https://www.-techniques-ingenieur.fr/base-documentaire/procedes-chimie-bio-agro-th2/innovations-en-genie-des-procedes-42487210/intensification-des-procedes-j7002/.

Poux, M. (2010). *Génie des procédés durables. Du concept à la concrétisation industrielle*. Dunod, Paris.

4

Operations

PRELIMINARY NOTE.– This chapter covers operations, also known as production or manufacturing. It is a key function and therefore a prerequisite for the survival of the industrial enterprise that ensures revenues through the proceeds of sales. Its primary objectives are: to serve customers, provide them with added value, and integrate into their value chain. The days when demand exceeded supply are now over. This was the case for France during the *Trentes Glorieuses* (1945–1975). Today, consumers find almost every product they need; and they have choices. One could almost say the whole world is at their doorstep! This is the case for the automotive, audiovisual, chemical and pharmacy sectors, to name but a few. Effectively managing the industrial tool, developing it, improving its performance, competitiveness, reducing costs, supplying at an ever-faster pace customers who have to be retained – all of this makes up permanent challenges for the manufacturer to consciously create jobs and wealth. Japan, a country with scarce resources, shaken almost daily by earthquakes that fortunately are most small in amplitude, has risen to second place in the world behind the United States, which is three times more populous. This has been possible by reconsidering the country's production methods during the post-war period when it was devastated. Toyotism has allowed Japan to become a major producer in the automotive industry, beating the Americans at their own game. Japan has been able to extend this set of techniques to other activities with the success that we know.

That is indicative of the importance of the design of the industrial tool and its management, which are intimately linked. Cleverly managed technology is the foundation itself of the company's know-how: it has become a weapon in today's global economic war.

Chapter written by Jean-Pierre DAL PONT.

This chapter by no means provides an exegesis of *everything* that is "industrial": this would take several volumes! We have tried to highlight key points, references that we hope will help the reader in their understanding of a primordial area and encourage them to further deepen it. Topics regarding scheduling, maintenance, safety aspects, profitability criteria, investment selection, etc. are not treated here or are mentioned only briefly. Readers will be able to take advantage of the articles and works mentioned in the references section, especially (Dal Pont 2012) and (Dal Pont and Azzaro-Pantel 2014).

4.1. The industrial tool seen by flows and Enterprise Resource Planning (ERP)

The industrial tool can be seen by its flows (Figure 4.1):

– Information flows: data concerning commercial orders and the state of the industrial tool have to be accessible. Is it capable of processing the order, both from a human resources and technology perspective?

– Product flows: raw materials, work in progress and finished products. Every product to be supplied, stored and handled.

– Financial flows: money has to be spent to buy raw materials, to pay for incurring fixed costs (see below).

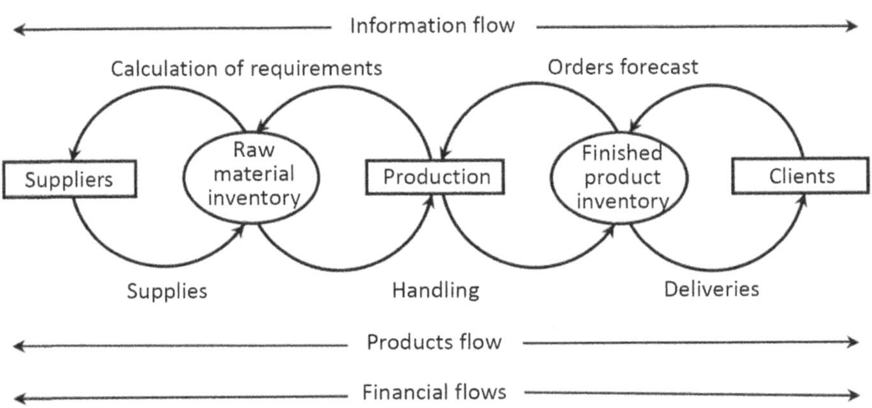

Figure 4.1. *The industrial tool seen through the flows*

ERP (Enterprise Resource Planning) refers to software packages that allow the management of all of these flows, including sales, inventory, accounting, planning and distribution. A well-designed ERP allows all players involved in the productive system (sellers, buyers, producers, logisticians, marketers) to have the same global vision at the same time. SAP (Systems, Applications and Products), of German origin, has emerged as a clear leader in this field, now unavoidable, which has replaced many school notebooks.

4.2. The supply chain

This concept was invented by Martin Christopher, author of *Logistics and Supply Chain Management* (2016). It brings together all the processes mentioned above along with an underlying concern for improvement and a desire to provide added value to the customer. It includes the concept of logistics, more focused on transport, storage and distribution, as illustrated in Figure 4.2.

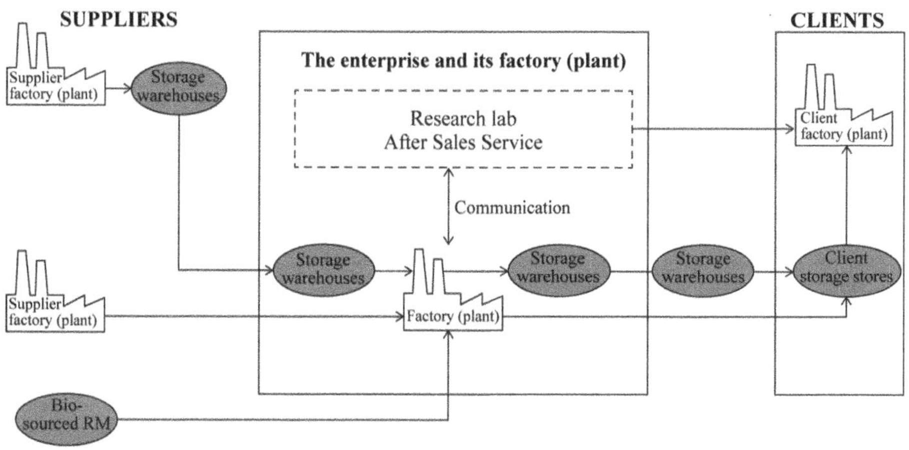

Figure 4.2. *Visualization of the supply chain (RM: raw materials)*

The complexity of the process is often unappreciated although the administrative aspects it underlies are apparent. **Traceability** has become an inescapable necessity in many cases (see Box 2.4, Volume 1). Bio-based raw materials of plant origin often raise particular problems due to their seasonality and the variability of their composition. This is the case for the sugar beet industry, whose plants operate only a few months a year: the roots must be sent to the plant quickly or risk losing sugar.

The concept of product stewardship (responsible product management) focuses on the product itself. It requires the producer to assume environmental responsibility and, in line with the above, to follow it until it reaches the end consumer. The producer must provide safety-related documents (SDS files), define the means to be implemented in the event of an accident during transport, etc.

4.3. The typology of the means of production: VAT analysis

The visualization of the product flows at an industrial site is important because it determines its internal logistics. When designing the site, it is a major factor in determining the layout or the implementation and the master plan. The latter takes into account the evolution of the site and its extension. The plant layout for its part, describes the connection between storages, warehouses, units, and all the services of the plant. Pipe racks are part of the industrial landscape.

The huge variety of production tools in the process industries has already been highlighted. Let us recall the very different characteristics between continuous and discontinuous production units (batch), as well as between units that handle fluids and those that handle solids. The dangerous nature of the products often gives the units that manufacture them a special character (see Box 3.2).

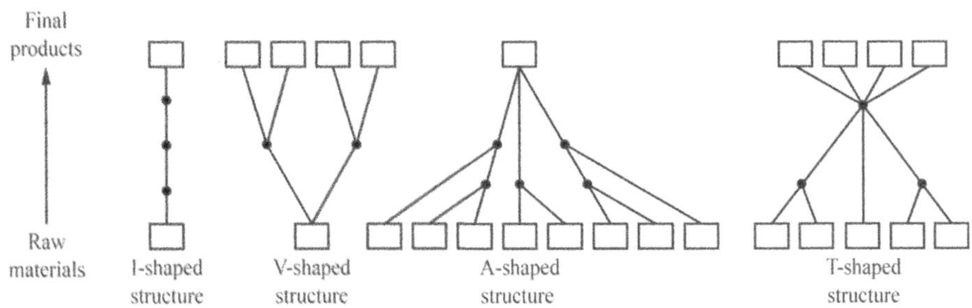

Figure 4.3. *VAT analysis*

Figure 4.3 illustrates what is known as VAT analysis, described in detail in (Dal Pont and Azzaro-Pantel 2014, p. 122 and onwards). The I-shaped structure is typical of chain production, that of the V-shape of the oil industry and the AI (agrifood industries): a product like maize generates a whole family of products. That of the A-shape, for example, relates to the assembly of an airplane: the different parts that it comprises are brought to the site. The T-shaped structure is typical of the fact that a product (masterbatch in the chemical industry) is diversified into several products: a laundry product is perfumed, colored, and packaged in different forms.

The VAT analysis will influence the choice of the ERP and the organization of the unit (the workstation and their ergonomics). The very small number of staff is striking when visiting very large oil facilities.

4.4. The anatomy of a plant

In Chapter 1 of the first volume, we sketched out the anatomy of a company. In the following, we shall do the same with respect to a plant. The latter can employ a few people or tens of thousands, not to mention outsourcing[1]. As already mentioned, the functions, according to Henri Fayol, can be grouped in the same department under the same umbrella: this has advantages and disadvantages. Creating borders and interfaces is often the source of difficulties.

We shall distinguish between:

1) the **plant administration**[2] and **supporting functions**[3]:

- director, secretariat. The plant manager is accountable to the law. He/she is the preferred interlocutor of headquarters;

- payroll, staff, accounting, IT, financial management, communication, HSE function (hygiene, safety, environment, which includes fire protection), quality, training, occupational medicine, infirmary, first aid, etc.[4];

2) **production**[5]:

- production management, chemical unit managers;

- supporting functions;

- logistics (procurement, storage, warehouses, maintenance, shipping and receiving), planning, technical assistance to units;

- engineering department;

3) **maintenance**:

- unit maintenance: central maintenance, on-site;

- the maintenance department may include; electricity, instrumentation, automation;

---

1 An organizational chart describes the organization of a plant in its hierarchical and functional aspects.

2 Generally, the plant is administered by a management committee composed of representatives of the key functions.

3 The functions of the plant are linked to the central functions of the company's headquarters.

4 The union(s) representation(s) may play a major role in the operation of the plant.

5 The factory can, in its operational mode, be the subject of major works: such as the construction of a new workshop, revamping (modification of an existing workshop, in principle to restore its original beauty). The project manager of this activity should generally be integrated into the management team.

4) **utilities, environmental protection**:

- utilities: steam, cooling water, compressed air, nitrogen, oxygen, brine, residue incineration, etc;

- environment protection; WWTU (wastewater treatment unit), waste treatment;

5) the **analysis laboratory**:

- control of raw materials, works in progress, products;

- claims management;

- possibly some basic research: testing of raw materials, production assistance;

6) **ancillary functions**: catering, guarding, security, hosting of visitors, organization of first aid.

4.5. Operations management systems, the push for excellence

A factory, as any organization, requires a **management system** (MS), to operate, that is understood and accepted by everyone. It must, of course, comply with a significant number of laws and regulations, in particular the labor code. This essential aspect – no one is supposed to ignore the law – falls outside the scope of this book. Each function of the plant must have a management system that corresponds to its activities. For example, the HSE department implements a security management system (SMS) – see for example (Charavel 2002).

Being a complex system, the plant must have a consistent system, shared, and based on **procedures** in the form of **written documents** describing the multitude of its processes.

4.5.1. *A brief history of industrial operations management*

Chapter 1 of the first volume mentioned the creation of the company at the interface of the 19th and 20th Centuries and its evolution up until the present day.

For the record, let us recall that Frederick Taylor was the first to take a critical look at the world of labor, which he recorded in his book *The Principles of Scientific Management* (1911).

The advent of computer science in the 1960s brought profound changes (see Chapter 5).

The 1980s saw the emergence of **management systems** affecting the modus operandi of the factory and the company as a whole. This was the case for safety. The **HAZOP** (Hazard and Operability Study) method was one of the first SMS, developed between 1960 and 1970 by the ICI Society and published in 1983. It was at this time that the French chemical industry implemented the HSE function (Royer 2016).

Quality methods also emerged in the late 1980s. Founded in 1988, the French Association for Quality Improvement and Management (AFAQ) merged in 2004 with the French Standardization Association (AFNOR).

Japanese methods, including **Toyotism**, began to also be known in the United States during this period. **Walter E. Deming**'s theories, back from Japan, gradually succeeded in establishing himself in his country: just a fair twist of fate! Deming contributed to the recovery of post-war Japan; considered the father of quality in this country, he did not receive any serious attention at the time in the United States. He believed that the American top management was suffering from "5 deadly diseases", and was therefore ineffective. Helped by the media, he stated his "14 points", which constitute a homogeneous package, and aim to reduce costs and create jobs.

The book by **Robert H. Hayes** and **Steven C. Wheelwright** (Hayes and Wheelwright 1984) is like a wake-up call. The authors draw parallels between Japanese companies and their American equivalents. The superiority of the latter relied on their "manufacturing superiority"[6] and their "ability to make it better"[7]. In short, the Americans neglected their production system; the authors believed that "manufacturing can be a formidable competitive weapon if equipped and managed properly."

The in-depth study of Toyotism – the Japanese began to manufacture in the United States – and the reflections issued from academic works in the wake of those that we have just been mentioned would lead to a number of notions in the form of acronyms that the present era loves using. Their descriptions can be found in (Dal Pont 2012) and (Dal Pont and Azzaro-Pantel 2014).

We shall quote:

– **WCM** (World Class Manufacturing) in which we give credit to Hayes and Wheelwright and which they describe extensively in (Hayes and Wheelwright 1984);

---

6 "Superiority of the production system."

7 "Ability to improve."

– **SQC** (Statistical Quality Control) which leads to **TQM** (Total Quality Management);

– **SPC** (Statistical Process Control);

– **TPM** (Total Productive Maintenance);

– **Six Sigma**: a method that aims to reduce product variability, thus to reduce "non-spec" products, namely products that do not meet the expected specifications and which will have to be recycled or destroyed along with the associated financial losses;

– **lean manufacturing**: this is a Western transposition of Toyotism. Lean implies fat-free production. This is to avoid waste (*muda* in Japanese) and to aim for zero defects.

4.5.2. *Toyotism*

In 1945, Japan was all but destroyed and the people were starving. The United States helped rebuild it under the leadership of General Douglas MacArthur. The latter called upon American experts, including W.E. Deming, which would contribute to the establishment of a quality system, knowing that Japan has its own system, the Kaizen, synonymous with continuous improvement. However, the Japanese miracle might be described in a single word: Toyotism, the basis of the empire that the Toyoda family would build.

The Toyoda family moved from pre-war looms to the automotive industry after the war; they created the Toyota Group. The success of the TPS (Toyota Production System) can be considered to have been driven by three components: a management system, the observation of people at work, and the implementation of management tools.

4.5.2.1. *Toyotism as a management system*

The engineer Taiichi Ono can be considered as the father of Toyotism as a system. After studying Fordism, he implemented JIT (*Just in Time*) manufacturing: flows were drawn, production was started when the order was available, stocks were eliminated and wastefulness was fought against.

4.5.2.2. *People at work*

The worker is involved in the study of workstations and in the decision-making process. We are far from Taylorism: it can be said that Toyotism is an enlightened Taylorism. Workers are provided with a number of tools to improve performance and assist them: 5S, Poka-Yoke, SMED, Kanban, visual management, etc.[8]. The operator is at the center of the unit system (shop floor).

4.5.2.3. *Management tools*

Management tools include: the house quality, statistical process control, problem-solving techniques (Ishikawa diagram), Quality Function Deployment, etc.

Figure 4.4. *A Toyota factory*

The question was raised as to whether or not the success of Toyotism was dependent on Japanese culture. Could it be exported? Toyota's success in France seems to confirm its universality: what could be more different from the French mindset than the Japanese mindset! Nonetheless, one should make no mistake. It took more than 10 years for Toyotism to take place in Japan. It even caused strikes! However, the strikers were still working; in protest, they wore a bandana around their forehead to show their disapproval.

4.6. Costing-based profitability analysis (CO-PA): measure of performance and steering tool

Readers are asked to refer to references (Dal Pont 2012) and (Dal Pont and Azzaro-Pantel 2014).

Typically, the manufacturer and engineer ought to know any costs, whatsoever: raw material costs, labor costs, maintenance costs. Manufacturers must know the

8 Explanations about these notions can be found in (Dal Pont 2012) and (Dal Pont and Azzaro-Pantel 2014).

cost price of the products from their units. They must also know enough accounting to be able to engage with managers, but they are not required to be accountants!

There is no management without measurement.

4.6.1. *Product cost*

The product cost is the sum of proportional costs (PC) or variable costs and of non-proportional costs (NPC) or fixed costs: Product Cost = PC + NPC.

4.6.1.1. *Proportional costs (PC)*

PCs include the costs of raw materials and utilities (UT) and are expressed in €/kg. They vary with tonnage, unlike NPCs which are fixed. PCs depend on the chemical route and the access price of raw materials. The cost of some strategic raw materials may depend on negotiations with the supplier. It can be a major variable in a project that will only succeed if there is agreement. An agreement with a supplier may originate the creation of a JV. PCs are, in principle, independent of the plant load.

4.6.1.2. *Non-proportional costs (NPC)*

Non-proportional costs of a product (NPC) include two components:

– direct **NPCs** that represent labor costs (LC), manufacturing (direct and supervising), maintenance and laboratory costs. *Direct NPCs are expenses that the new unit created in the factory that hosted it*;

– indirect **NPCs** or **PSCs** (plant specific costs).

A unit that is located in an existing factory will benefit from its infrastructure.

In summary:

– administration: management, staff service, accounting, guarding, etc.;

– support functions: fire department, security department, infirmary, utilities production, WWTU, road maintenance, etc.

Indirect NPCs (indirect fixed costs) are in a sense the *housing tax* that the product pays to the hosting plant. They are *broken down* according to products using *keys* defined with the accounting department. For example, fees related to administration functions may be broken down proportionally to unit staff.

For the same product, direct NPCs (direct fixed costs) will vary depending on the size of the unit, operation rate (percentage of plant load), the size and organization of the host site. Unlike PCs, NPCs are not easily accessible; they will depend on ramping up of the facility. *The cost of a product decreases as the volume of sales increases given that the fixed costs will be broken down across a larger tonnage.*

4.6.2. *Margins*

Figure 4.5 illustrates costing-based profitability analysis (CO-PA). It is an essential management tool. It can be applied to a product, a family of products, at the plant or the company level.

– The contribution margin represents the selling price minus proportional costs. This margin covers the fixed costs of the product. At the factory level, the contribution originating from the products covers its fixed costs.

– Gross margin (*gross profit*) represents the selling price minus PC minus NPCs. This is the margin of the product within the plant perimeter.

– The industrialist must also pay Head Quarters Costs (HQC) (building costs, administration, R&D, commercial costs). He/she must also pay financial costs (borrowing) and taxes if profit has been generetated.

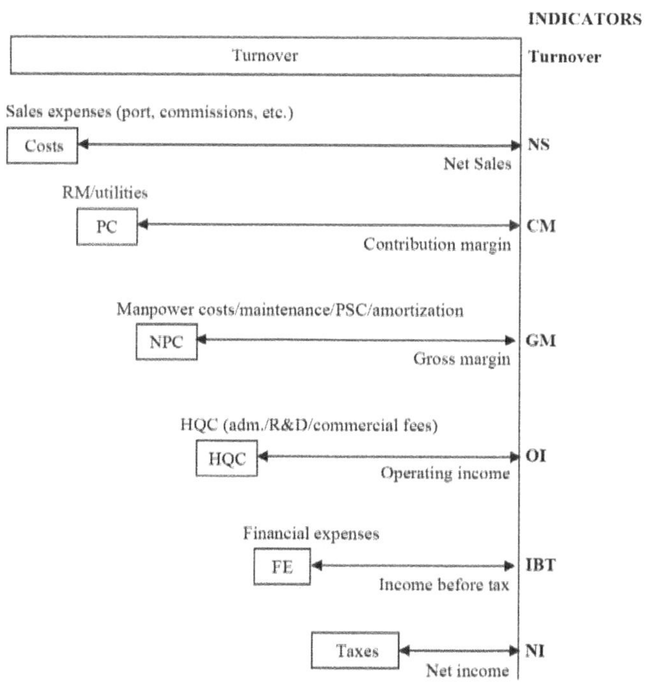

Figure 4.5. *CO-PA*

4.6.3. *The breakeven point: the absorption of fixed costs*

Figure 4.6 highlights the correlation between the contribution margin and fixed costs. The breakeven point is the **minimal** tonnage for which the contribution of the product covers the fixed costs of the plant. The latter does not lose money, but does not earn any!

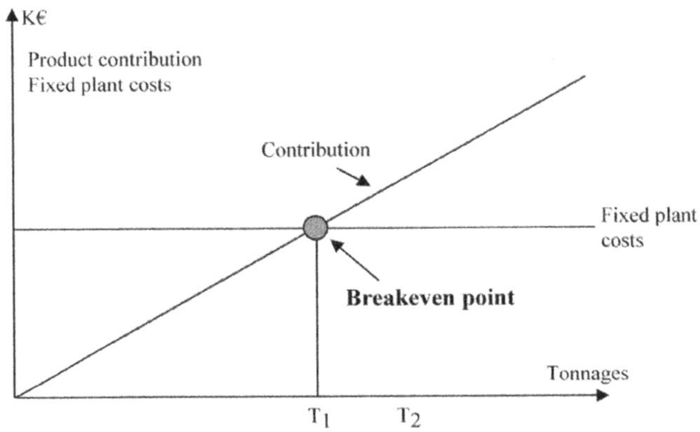

Figure 4.6. *The plant breakeven point*

4.6.4. *The infernal spiral of fixed costs*

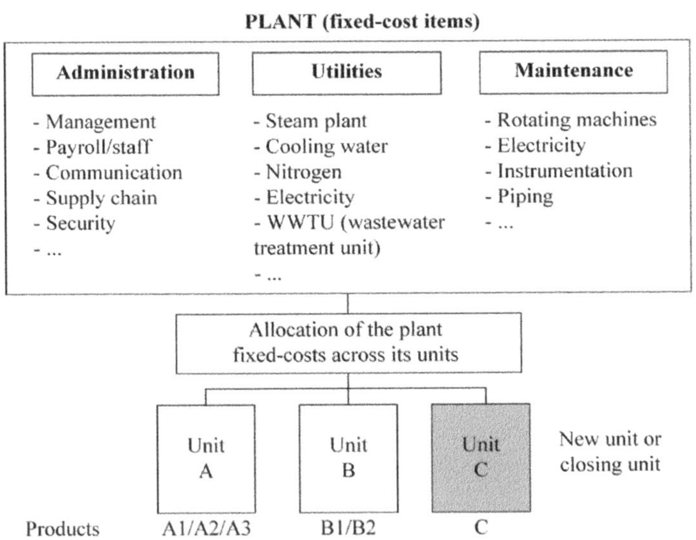

Figure 4.7. *Decomposition of the fixed costs of the plant across its products*

A factory-(plant), usually multi-product, allocates its fixed costs (referred to as indirect NPCs above) across its products according to keys (allocation keys) (see Figure 4.7). The creation of a new workshop (process unit) C in a plant which encompasses process units A and B will absorb part of the plant total fixed costs. Therefore, the cost of products A and B will decrease although the plant fixed costs will increase, given that workshop A will create its own direct fixed costs. Conversely, if the unit C ceases operations, the fixed costs of the remaining workshops A and B will increase. When France began its deindustrialization, this perverse effect, causing a snowball effect, sometimes referred to as the fixed costs' infernal spiral, where the closure of factories led to the closure of other factories.

4.6.5. *Observations on margins*

Any product must generate a minimal contribution margin below which it is not profitable, since it is this margin that will cover downstream costs. To give an idea, it is about 40% for chemistry and 80% for pharmacy. This difference can be explained by the very high costs of research in the pharmaceutical industry. The chemist in charge of the research must calculate the contribution margin at the beginning of the study, as far as practical. Yields play a key role in chemistry. For example, we present in Figure 4.8 the variation in ortho-xylene consumption according to the yield in the case of its transformation in phtalic anhydride.

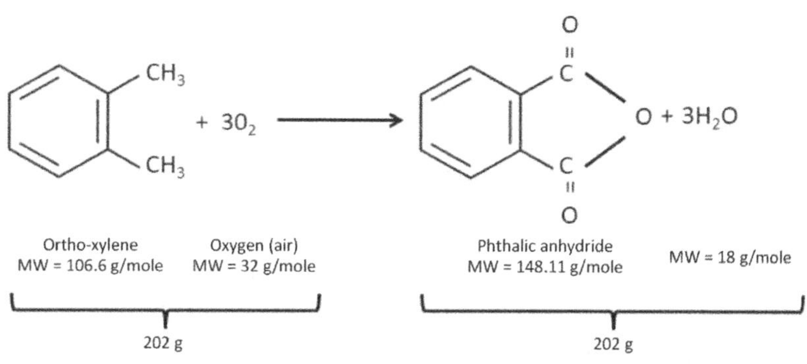

Reaction yield	Ortho-xylene consumption per gram of phthalic anhydride
100% (Stoichiometry)	0.719
90%	0.798
70%	1.027

Figure 4.8. *The importance of raw material yields*

A good **contribution margin** may depend on the purchase conditions of raw materials (the company has a good buyer). In the chemical world, one begins with cheap raw materials, then a good yield is obtained (providing the company has a good chemist). The gross margin depends on the size of the unit, its design, the type of management, the cost of the workforce and its versatility. A unit that is too small may notice that its cost of production will be burdened by its fixed costs compared to a larger unit.

4.7. The plant: performance measurement and score cards

A plant must measure its performance and inform stakeholders wisely as far as they are concerned. *A priori*, a human resources manager does not need to know a reaction yield. This activity introduces score cards, as a prerequisite for progress.

An accident, damage to the environment, or a human problem are the subject of a special procedure that can lead to a *crisis cell*, which must be the focus of an appropriate procedure. In other words, one should to be prepared just in case.

The following is given as an indication. Each organization must adapt its ***reporting*** and its indicators (KPIs, **key performance indicators**) to its activities.

The **executive summary** summarizes the situation in a few lines as recommended for project management:

– human aspects provide the *social temperature*:

- HSE, safety, accident frequency rate;

- absenteeism, hiring, resignation, dismissal;

- social conflicts;

- in a way, everything related to governance, that is, respect for the ethics of the company;

- relations with local residents and the surrounding community;

– financial aspects:

- budget versus real expenses for the month and on-going (YTD, Year To Date = at the present date);

- these may be the elements developed for the CO-PA (see above), therefore margins;

- major expenditures: labor costs, maintenance, key utilities (electricity, gas), major purchases;

- inventory situation;

– commercial aspects:

- deliveries;

- claims rate;

- OTIF (On Time In Full), deliveries made on time;

- supplier problems;

– production aspects:

- major problems;

- major specific consumptions: kg of raw materials/kg of finished product;

- HSE aspects in addition to safety, everything related to the environment;

- quality and waste;

- discharges, effluents (greenhouse gases (GHG), COD (chemical oxygen demand), polluted water);

- tonnages and equipment occupancy rates (OEE, Overall Equipment Effectiveness = the usage rate of the industrial tool);

- proportional costs (variable costs) analysis: variance due to cost of raw materials, variance due to difference in yield.

4.8. Change management

An organism that does not develop is likely to disappear: "To be inert is to be beaten" (Charles de Gaulle, quoted in (Le Monde 2020)). The company must develop, adapt to a changing environment, propose new products, find new customers, improve its services. The same is true at the plant level with a very diverse staff that ranges from white collar to blue collar.

At the plant level, the processes implemented are the ones that need the greatest attention. The plant should be seen as a system to which it is necessary to bring change. Change management can be considered a management discipline. As far as a factory is concerned, it is necessary to **cooperate** with the people in place, because, on the one hand, they have the know-how and, on the other hand, they cannot be replaced with a magic wand.

Change management has two inseparable components: one technological and the other human. It can lead to anxiety, fear, demotivation, animosity, rejection and conflict. The first step is to "thaw" a situation, an existing state of mind (*unfreezing*) and to overcome inertia. The second step is the change stage itself. In the third and final stage, a new state of mind takes place, that of *refreezing*.

We shall merely consider a few fundamental concepts that can be found in (Dal Pont 2012, p. 437 and further).

4.8.1. *Processes: system integrity and robustness*

An organization encompasses a multitude of **processes**. A process is a defined activity with a beginning and an end and has in principal a defined goal: for example, to hire an employee. This process can be divided into an important subset of other processes. In this case: process of expression of the requirement, announcement of the requirement, candidate(s) interview, final selection. Process analysis is time-consuming, which explains why people resort to outside companies.

An industrial process is usually described by a process book, which is an essential document that describes the process: reactions, normal operation, start-up, shutdown of the facility, etc. It is a set of procedures. A **procedure** is a process put in written form (SOP, Standard Operating Procedure).

Any change has an impact. It is expected that it is favorable, which is not always the case because of a lack of knowledge of the field. A new raw material purchased at a price below the usual price can lead to setbacks. For example, purchasing low cost coal containing silica in small quantities resulted in exorbitant maintenance costs due to the abrasion of economizers.

It is crucial to preserve the **process in its entirety** and any changes must be validated by a process similar to that of project validation (see Chapter 3). The system must preserve its **robustness** and its **resilience**.

4.8.2. *Human aspects and climate of trust*

The chances of success during change depend on people (*people make things happen*). It seems essential:

– to state the rationale, the objective of the change itself, especially if it is tedious;

– to quantify the expected improvement and implement adequate reporting;

– to make the system visible: visual management is defined as the set of measures that allow all employees to understand the system to which they belong. This ranges from signage, the standardized colors of pipes, and billboards, all the way up to procedures known to everyone, to the safety, quality system, etc. We are talking about a transparent factory;

– to determine who is responsible for whom and what. Experience shows that many people find it difficult to integrate themselves in the company, as shown in Figure 4.9. This can lead to burn-out, depression, or worse.

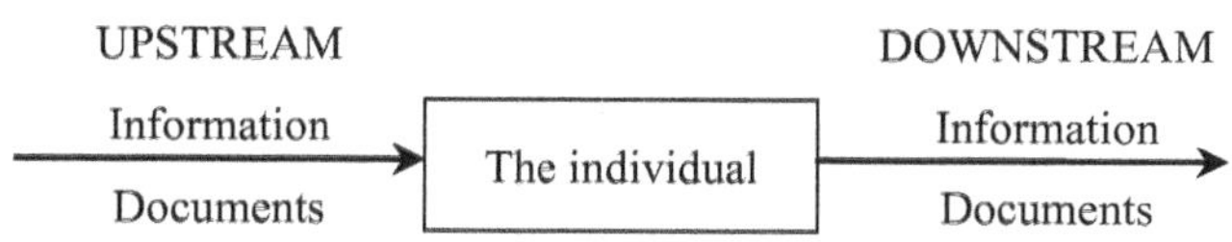

Figure 4.9. *Individuals and their integration in an organization*

This minimalist figure expresses an extremely simple fact, though one that is crucial. Every person has an "upstream and a downstream"; he/she expects information and documents in general such as meeting reports, meetings, or mail of all kind from the "upstream". His/her downstream awaits just as much. Every individual needs recognition.

4.8.3. *Knowledge management and core competencies*

The company's first capital is its know-how and therefore its **human capital**. Our societies have become societies based on science and technology. It is essential if we want to remain in the top league.

Knowledge management, or KM as it is commonly referred as, concerns practices that aim to define the core competencies of the company. In other words, what knowledge does the company need to manage and develop its technology? What skills should be bought (initial training)? What continuing training should be provided to the employees?

Defining these core competencies is a difficult exercise. Every technician should consider whether his/her **toolbox** allow him/her to remain competent and, as such, competitive within the labor market. What skills should be developed, or acquired?

4.8.4. *Continuous improvement and the search for innovation*

Any organization can and must improve. It can do so with the means available, in other words with the team in place. Figure 4.10 describes the mechanism: this is the Deming wheel.

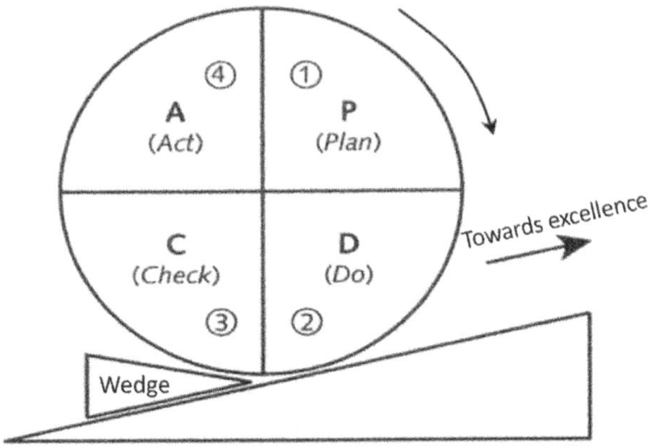

Figure 4.10. *The Deming wheel, continuous improvement*

The four quadrants of the wheel describe the four aspects of the process: PDCA (*Plan, Do, Check, Act*). The wheel is on an inclined plane, bearing witness to the fact that any change is difficult and requires commitment and consistency. The wedge that prevents the wheel from receding symbolizes the quality management system. Progress has been achieved; the system is "frozen". Then we continue: P, D, C, A!

4.8.5. *The search for technological breakthrough and innovation*

It is important for the company to know if its technology is mature, if it is likely to become outdated or if it is already so.

In the case of a unit, the company may decide to examine an existing process in depth. To do this it will set up what is called a PIT: **a process improvement team. The goal is to create a technological breakthrough, to seek innovation**.

Innovation differs from invention, which is a discovery due to the fact that it has commercial benefits.

Management commitment is a prerequisite for success. It is evidenced by the release of funds and its interest in monitoring this activity. How many projects never get off the ground!

Some tips	Mistakes to avoid
– A multidisciplinary team of 5 to 7 people from outside and inside the plant proficient in the field under study. – An external leader recognized and supported by top management. – Allocated resources (lab studies, engineering). – A budget. – A well-identified methodology. – Short-term, long-term challenges displayed (Pareto analysis). – Measured, monitored and coordinated progress. – Recognition.	– Non-available, unrecognized people. A non-specialist leader in the field. – Inadequate means. – Lack of continuity. – Lack of interest from top management. – No budget allocated.

Table 4.1. *The implementation of a PIT*

4.8.6. *Operations abroad*

Any company that has operations abroad wants to ensure the cohesion of its operations and above all respect for its governance. Normally, its management system is universal, but remoteness, cultural shock, and language problems should be taken into account. Mailing, conference calls, and video conferencing have changed a lot of things.

Nevertheless, local visits, verifications if not audits, are to be scheduled. The parent company must invite its foreign workforce; it creates a sense of belonging.

4.8.7. *What about tomorrow?*

The digital revolution is leading to an unprecedented, out of the ordinary potential for change. It overturns old habits and involves training needs in new areas.

We believe that the above applies. "And tomorrow..." is the title of the last chapter of this book. The reader will be the judge of that.

4.9. References

Charavel, B. (2002). Système de management de la sécurité : mise en place sur site. Techniques de l'Ingénieur [Online]. Available at: https://www.techniques-ingenieur.fr/base-documentaire/procedes-chimie-bio-agro-th2/industrialisation-des-procedes-et-usine-du-futur-42602210/systeme-de-management-de-la-securite-mise-en-place-sur-site-ag4650/.

Dal Pont, J.-P. (2012). *Process Engineering and Industrial Management*. ISTE Ltd, London and Wiley, New York.

Dal Pont, J.-P., Azzaro-Pantel, C. (2014). *New Approaches to the Process Industries*. ISTE Ltd, London and Wiley, New York.

Hayes, R.H., Wheelwright, S.C. (1984). *Restoring our Competitive Edge: Competing Through Manufacturing*. Wiley, New York.

Le Monde (2020). Etre inerte, c'est être [...] – Charles de Gaulle [Online]. Available at: https://dicocitations.lemonde.fr/citations/citation-15951.php.

Royer, M. (2016). HAZOP : une méthode d'analyse des risques – présentation et contexte. *Techniques de l'Ingénieur* [Online]. Available at: https://www.techniques-ingenieur.fr/base-documentaire/environnement-securite-th5/methodes-d-analyse-des-risques-42155210/hazop-une-methode-d-analyse-des-risques-se4030/.

5

The Enterprise and the Plant of the Future at the Age of the Transition to Digital Technology

What is commonly known as the transition to digital technology, which we are experiencing today, succeeds to the ecological transition and the energy transition mentioned in Chapter 2 of the first volume. It is the culmination of unparalleled changes in the history of humankind. Let us remember that Napoleon's armies were moving at the speed of Julius Caesar's!

We are talking about revolutions, we shall describe them succinctly.

5.1. From one Industrial Revolution to the next Industrial Revolution

The current Industrial Revolution, the Fourth, is part of a long line of technological advances, breakthroughs, new types of organization and a wide range of political systems from royalty to republics, not forgetting dictatorships, economic crises and revolts (Dal Pont 2012).

The Industrial Revolution cannot be separated from the social system (Dal Pont 2012) and from the freedom granted to people, to entrepreneurs to invest in companies that involve capital, and therefore risks. Capitalism in the financial sense plays a fundamental role. It is not always well understood to this day. Joyce Appleby in her masterful book *The Relentless Revolution* (Appleby 2010), defines it as:

> A system based on individual investments to create marketable goods.

Chapter written by Jean-Pierre DAL PONT.

Britain left its mark on the 18th and 19th Centuries, followed in this movement by France, Germany and other European countries, then by Japan and the United States, whose power dominated the world of the 20th Century. Today, developed countries are engaged in a frantic **economic war** whose technology is the essential weapon and where battles are financial. No one can say where this will lead us. **The only certainty is uncertainty.**

5.1.1. *The First Industrial Revolution (1712–1860): steam, a source of energy*

The exploitation of iron and coal mines would not have been possible without the pump of the British blacksmith Newcomen (1663–1729) (Dal Pont 2012, p. 15), one of the first "fire machines" to extract water that floods mines (see Figure 5.1).

Figure 5.1. *Newcomen's pump*

Inventions are at the root of the phenomenal development of the textile industry and the steel industry, which turned Britain into the industrial country we understand it to be today.

The figure of James Watt (1736–1819) dominated the whole of the 18th Century: he improved Newcomen's machine into an efficient steam engine without which the Industrial Revolution would not have had the magnitude that it had. His association with Matthew Bolton has remained famous: it is that of a brilliant engineer and an inventor, who was a very good manager since he knew how to handle capital and allowed his partner to deal with technical issues.

Stephenson's locomotive of 1815 was at the origin of the development of the railways, fostering unprecedented economic and technological exchanges.

This period ends around 1850–1880 depending on the country. The American Civil War (the War of Secession, 1861–1865) was the first technological war in which all inventions were implemented to annihilate the enemy. The post-war period saw the unprecedented rise of the United States.

5.1.2. *The Second Industrial Revolution (1860–1960): from crafts to industrial enterprise*

5.1.2.1. *Humans, robots, machines*

This period spanned a hundred years and ended around the 1960s with the advent of computer science. It was characterized by:

– the "Europeanization" of the planet and the colonization that created the so-called colonial wars after the Second World War;

– the weakening of imperial China as a result of the unequal treaties in 1850 that ended the Opium Wars;

– the founding of the People's Republic of China (PRC) in 1949;

– two world wars that brought Europe to its knees;

– the increase in the world's population from one billion individuals in 1830 to two billion in 1930;

– inventions that followed one after another at an ever-increasing rate. To name but a few:

- irreplaceable electricity brings power to electric motors to activate diverse machines. Trees, pulleys and belts disappeared from factories,

- steel replaced iron,

- the phone brought people together,

- chemistry developed dyes, synthetic fibers (nylon), fertilizers, medicines including the famous penicillin,

- oil, which was easy to transport, allowed the growth of the automotive industry, synonymous with mobility.

Figure 5.2. *A workshop with pulleys (Pulleys and belts, Escher Wyss, 1875)*

The American industry was the "arsenal of democracy" during World War II, to use President Roosevelt's phrase. The United States was providing weapons and equipment to the Allies, including Russia. Their aviation and navy were unmatched.

The first application of nuclear energy saw the construction of atomic bombs. The first was dropped over Hiroshima on August 6, 1945. Nuclear power plants would follow.

After the Second World War, the consumer society found itself enjoying household appliances, televisions and cars.

The explosion of the first Russian atomic bomb in 1949 saw the United States relaunch the arms race. The two blocs clashed during the Cold War, which ended with the collapse of the USSR in 1991.

This post-war period gave rise to a reflection on the human–machine relationship (cybernetics), the development of robots and the birth of the computer. Machinism, defined as the use of machines that substitute for a person, has been the object of reflections that persist to this day.

Capital, which allows the purchase of machines and "enslaving" those who make them work for the benefit of the owner, fueled the reflections of the socialists of the 19th-Century. Karl Marx (Attali 2005) calls "goods" the labor products that have "use value", "exchange value" and a price. Workers only sell their "labor force" for a salary to the boss who creates "goodwill" and therefore profit because of the machines that he/she alone owns (*The Capital*).

In 1801, Joseph-Marie Jacquard (1752–1834) invented the first programmable loom using perforated cards. His invention, like others, led to a social revolution. Silk workers saw that the workforce would be reduced (only one worker was needed to operate the Jacquard machine) and feared unemployment. They went after its inventor, who almost ended up in the Saône river.

5.1.2.2. *Cybernetics*

Norbert Wiener (1894–1964) was the inventor of the term "cybernetics", which was in fact introduced by André-Marie Ampère. Wiener, MIT's leading mathematician, worked on the autopilot system of anti-aircraft guns during the war. His 1948 book, *Cybernetics*, propelled him to the forefront.

Being an information science, cybernetics is an extension of the concept of a system as envisioned by Von Bertalanffy (Dal Pont 2012, p. 447). Its main focus is on the interaction of the elements between one another, introducing the notion of feedback, to the very basis of system control. The concept of the human–machine relationship is studied in particular. In his book *Rise of the Machines* (Rid 2016), Thomas Rid cites the pilot-aircraft as an example. Pilots needs to communicate with their aircraft to make it fly; the aircraft needs to communicate with pilots to fly.

The fear of Russian aggression led the United States in the 1950s to equip itself with a system composed of radars weighing hundreds of tons connected by telephone cables: this was the beginning of networks. The point was to detect approaching enemy aircrafts and to destroy them. This is what would be soon nicknamed the push-button war.

A cyberculture develops robots, imagines thinking machines; it then becomes difficult to discern what is real and what is mythical or fantasy. The term "cyborg" is the contraction of "cybernetic organism" (Rid 2016, p. 113). The first cyborg was a rat to which a pump implanted in the tail delivered chemicals (Rid 2016, p. 125).

The objective was to study how to increase capabilities in humans so that they could survive in space.

The cyberculture trend does not survive Norbert Wiener. Only the terms "cybercrime" and "cyberattack" are used today.

5.1.2.3. *Computers and humans*

Humans have long sought to facilitate computational operations: the abacus is still used today. A young 19-year-old Blaise Pascal invented the pascaline in 1642 to help his father with his calculations (see Figure 5.3).

Figure 5.3. *The pascaline (David Monniaux,* Musée des arts et métiers, *Paris). For a color version of this figure, see www.iste.co.uk/dalpont/process2.zip*

Early on, humans also tried to control the machines in order to perform specific tasks. The history of computer science is full of countless events and inventions. Here are just a few examples of benchmarks:

– Alan Turing (1912–1954), a brilliant British mathematician, was the forerunner of software. His computers, curiously referred to as "bombs", broke the code of Enigma machines used by Nazi Germany, making a considerable contribution to the war effort, especially during the Battle of the Atlantic. It laid the groundwork for AI (artificial intelligence) (Copeland 2012);

– John von Newman helped Pres Eckert and John Mauchly design and build the ENIAC (Electronic Numerical Integrator and Computer) between 1943 and 1944. The goal was to develop calculation tables for firing cannons. What can be considered as the first computer uses 5,000 to 18,000 cathode tubes and weighs 30 tons. The modified ENIAC led to the EDVAC (Electronic Discrete Variable Calculator), which was marketed around 1952 (McCartney 1999);

– the IBM 360 computer was the first non-military computer to be introduced in the 1960s. It created a real revolution for the big companies that could afford it.

5.1.3. *The Third Industrial Revolution (1960–1990): the rise of industrial computing*

5.1.3.1. *The beginnings*

– 1876: Alexander Graham Bell invented the phone.

– 1906: people's voices were broadcasted by radio in the United States.

– 1926: first TV show.

– 1948: first transistor.

– 1957: Sputnik 1, the first artificial satellite, triggered the Space Race.

– 1958: the PERT (Program Evaluation and Review Technique) diagram created by the U.S. Navy made it possible to save two years or work via better planning the construction tasks of POLARIS nuclear warhead rockets.

5.1.3.2. *The period 1960–1990*

– 1969: Neil Armstrong set foot on the Moon on July 21 (Apollo 11 mission).

– 1971: development of the microprocessor.

– 1971: 20 computers were connected on Arpanet. This was the beginning of the Internet.

– 1980: the personal computer invaded homes.

– 1983: Motorola's first mobile phone.

– 1984: Apple launched the Macintosh.

– 1990: Internet.

– 2000: the smartphone appeared. The "internet bubble" exploded.

In the 1980s, Steven Wozniak and Steve Jobs, aged in their twenties, founded Apple in a garage. Michael Dell was building computers in his student room (Appleby 2010, p. 343). In 1984, Apple's Macintosh was a huge success; it featured a mouse.

The 1990s saw the rise of the World Wide Web (WWW) at a staggering pace. The Internet allowed millions of people, and then billions, to have access to all kinds of information, to make phone calls, and to watch videos. The number of connected sites reached millions. The increased power of computers would create a revolution that we are experiencing today.

5.1.4. *The Fourth Industrial Revolution (1990–present)*

This Fourth Industrial Revolution originates from the **increased power** of computers that follow Moore's law stipulating that their power doubles every 18 months. It is characterized by the following facts:

– their decreasing cost, their ease of use, their increasing number;

– the advent of smartphones and tablets;

– the growing number of Internet users and the development of social networks;

– the emerging of increasingly sophisticated, miniaturized sensors and tracers, whose cost decreases over the years, and allows all systems to be monitored in real time. The Global Positioning System (GPS) makes it possible to locate everything, including oneself. It has been used both for the launch of the Millau viaduct and for locating truck fleets. Chemical analyzers have their detection limits reaching unimaginable thresholds, which means that everything is found in everything, including pesticides in every food product. The miniaturization of analysis devices, described as **nomadism**, allows customs officers to detect dangerous substances using Raman hand-held devices. Raman devices, which were initially very large, have shrunk in size as they become more and more efficient;

– the use of satellites for communications and Earth surveillance, to which drones should be added.

This whole system for data capture is the basis of the Internet of Things (IoT), a revolution in itself, made possible by computer networks. This will be discussed later (see Boxes 5.4 and 5.8).

There are more and more connections between the **physical world** and the **digital world**. We talk about physical trade (we go to the stores) and e-commerce (we order via the Internet, even if it means that the refrigerator is capable of placing orders itself if an ingredient runs out. A drone can deliver it knowing that the customer's account will be debited in real time).

We are witnessing a dematerialization of exchanges that take place over the Internet preferably over paper (mail, billing, etc.).

The explosion of the "internet bubble" in March 2000 reflected the frenzy that was taking hold of the computer world at that time. A multitude of companies were marketing hardware and software products. Netscape Communications, on the day of its initial public offering (IPO) along with the launch of its browser of the same name, saw its stock triple and its market value reach 2.8 billion dollars at closing time! This explosion led to the disappearance of many companies, countless bankruptcies and caused the global stock market system to falter. Gafa (Google, Amazon, Facebook and Apple) were shaken.

Business, **capital** and **technology** are **closely intertwined**, in an increasingly unpredictable way and at an unprecedented speed. The globalization of information goes hand in hand with globalization: planet Earth becomes a village. There are no more barriers to communication!

5.1.4.1. *The ubiquitous connection*

The market capitalization of the American IT companies, known as GAFA, to which Microsoft and China's Huawei should now be added, is a worrying issue. This will be covered in more detail in Chapter 6.

The term "artificial intelligence" alone testifies to the ongoing Industrial Revolution, the changes in our lives, to our jobs, the way we move, the way we take care of ourselves. The term is invading books, articles, speeches, private and government research programs. It is claimed by companies of all sizes and by the many start-ups that arise therefrom. The boxes in appendices to this chapter illustrate the many facets of AI applications.

5.2. Artificial intelligence (AI): deep learning and machine learning

Unambiguously, "artificial intelligence does not exist", writes Luc Julia, one of the inventors of Siri, Apple's voice assistant, in his book *There Is No Such Thing As Artificial Intelligence* (Julia 2020). During his long sojourn in Silicon Valley, he proposed another name for it: augmented intelligence. He adds: "AI is programmed for a specific purpose, it will never do but what we ask it to do and learn to do."

In his book *L'intelligence artificielle expliquée à mon boss* (*Artificial Intelligence Explained to My Boss*) (Blanc 2018), Pierre Blanc points out that underlying the terms there is *information technology* and *algorithm* (p. 53). In his mission report to the French government "*Donner un sens à l'intelligence artificielle*" (Giving meaning to artificial intelligence) (Villani 2018), Cédric Villani highlights the difficulty of defining its ambitious content: "To understand how human cognition works and replicating it; create cognitive processes comparable to those of human beings."

Cognition can be defined as a set of faculties and phenomena that concern the human mind and its functioning: memory, attention, reasoning understanding, perception, etc. These mental faculties are essentially based on the brain consisting of neurons and glial cells (Collins *et al.* 2018). Humans have 100 billion neurons connected to each other by synapses. They react to the stimuli of dendrites that cause them to send an electrical signal via their axons. Each neuron is connected to 7,000 neighboring neurons! AI is part of the cognitive sciences.

The term AI would have been first used at the conference that was held at Dartmouth College in Hanover (USA, New Hampshire) in 1956. The scientists that were present addressed the issue of whether a machine could be endowed with intelligence and perform tasks previously reserved for humans. Alan Turing in 1950 had wondered if a machine could think: didn't his "bombs" decrypt the German Enigma code during the war?

The fact that IBM's Deep Blue computer beat chess champion Garry Kasparov in a second confrontation in 1997 in New York made headlines: here was a machine smarter than a human being! AlphaGo beat Go world champion Lee Sedol of South Korea in 2016, the Go game being considered to be even more complex than chess.

Nevertheless, Pierre Blanc (Blanc 2018, p. 80) further clarified the idea: it was not a machine that beat Lee Sedol, but a hundred programmers who had used many machine learning algorithms. Machines are not able to contextualize. AI without humans is just a cold machine made of electrical circuits, a computer manipulation of 0s and 1s through algorithms (Blanc 2018, p. 44). For Pierre Blanc, a machine is synonymous with calculators, computers and robots (Blanc 2018, p. 29). Myths and fantasies created around AI seem to be fading, but others of a different nature appear; we shall discuss this in the next chapter.

An **algorithm** is a succession of instructions defined by a programmer and executed by a computer aiming to solve a problem or perform a calculation; the term comes from the name of the Persian mathematician Al-Khawarizmi, who lived in the 9th Century. A cookbook could be considered as a compilation of algorithms! Algorithmics is a branch of computing that studies algorithms.

AI, born from IT, is everywhere, without everyday people on the street even noticing it: weather prediction, facial identification, train schedules, autonomous cars, text recognition, machine translation, or chatbots. Chatbots use AI to answer user questions; store opening hours, parcel tracking, etc. They can give the illusion at least for a while that we are dealing with a human being.

AI is based on machine learning, which itself includes deep learning. Machine learning is a branch of AI that allows computers to learn from data without being programmed to do so (CEP Magazine 2018). **Deep learning** utilizes algorithms called artificial neural networks (CEP Magazine 2018). These networks mimic the functioning of human neurons. At its core, the perceptron is a mathematical model of the human neuron; it receives an input with an associated weight and delivers an output. It can be taught to separate images of cats from those of dogs. The arrangement of neurons in multiple layers connected to each other makes it possible to increase the accuracy of the expected result. Thousands of pictures of dogs and cats feed the machine, teaching it to distinguish a dog from a cat when a new image is submitted to it. **The more a neural network is used, the more it learns!**

AI relies on the huge amount of data from the Internet (big data) that the extraordinary power of modern computing is capable of processing and on increasingly sophisticated algorithms based on deep learning.

Box 5.1 by Philippe Chrétien is dedicated to the autonomous car and delineates the respective fields, the interaction of AI and human intelligence. The complexity of the system is brought forward. It takes hundreds of millions of programming lines for the car to be almost completely autonomous!

5.3. Big Data

For millennia, rulers have needed to store information about populations, wealth, number of herds, etc. The code of the Babylonian king Hammurabi (circa 1750 BC) engraved on a stele outlines the laws in force. Mesopotamian tablets and Inca quipus store transactions and historical facts.

a) b)

Figure 5.4. *Civilization and information storage: a) Mesopotamian clay tablet (source: Wikipedia, Jastrow, 2006); b) Inca quipu (source: Meyers Konversations-Lexicon)*

Today, the world is becoming numerical and the volume of data, whether related to society, business transactions, e-mails, social networks, scientific exchanges or transport, is staggering. Data production is currently estimated to be 3 trillion bytes per year (3 x 10^{18}) (Espinasse 2018). Facebook hosts around 100 billion photos a year! A byte is an 8-bit unit. A bit can take the value of 0 or 1 (bit is the contraction of binary digit).

1 kilobyte (KB)	= 10^3 bytes
1 megabyte (MB)	= 10^6 bytes
1 gigabyte (GB)	= 10^9 bytes
1 terabyte (TO)	= 10^{12} bytes
1 petabyte (PB)	= 10^{15} bytes
1 exabyte (EB)	= 10^{18} bytes
1 zettabyte (ZB)	= 10^{21} bytes

Table 5.1. *Number of bytes*

These data are only valuable if they can be stored, sorted and analyzed. Data are stored in files. The Web stores files worldwide. Open Data are data left free to users; they may be the work of communities (town hall meetings), businesses (train timetables) or research organizations. In 2009, in true democratic spirit, the United States launched the portal data.gov.

5.3.1. *Characterization*

We went from the 3V to the 5V model:

– **volume**: a notion that speaks for itself;

– **variety**: the heterogeneity of the sources has been reported. An industrialist is not necessarily interested in astrophysics data at first;

– **velocity**: refers to the time it takes to process information. It should be immediate in the case of data issued from remote monitoring systems;

– **value**: data acquisition has a cost. Are the stakes too high? Some old data, such as the work orders relating to the maintenance of specific equipment, are handwritten: digitization is necessary;

– **validity or veracity**: refers to the trust that is put in data.

A new profession has emerged: the **data scientist**, the person qualified for this new discipline.

Cloud computing or the cloud refers to a number of servers available to customers who cannot invest in increasingly gigantic computing means. Big Data are usually stored in the cloud. The customer therefore uses computing power made available to him/her via the Internet. **Cloud storage** can be seen as part of cloud computing as a service delivered to customers (see Box 5.6).

Big Data analytics implements algorithms, especially those of machine learning. It makes use of what is called **data mining**. It can be applied to data streams, in particular to continuously exchanged data, to text analysis, for example, including the automatic extraction of specific information. The Internet of Things (IoT) is an important source of data through its sensors and surveillance cameras, whether it is controlling an industrial process, a sports stadium or a fleet of vehicles.

Each activity requires specific algorithms. Opinion analysis is of course of interest to policy and marketing people.

Data are considered a source of wealth.

In the first part of his report "*Une politique économique articulée autour de la donnée*" (A data-driven economic policy) (Villani 2018), Cédric Villani recalls that "data are the raw material of AI on which the emergence of many uses and applications depends." He points out the colossal investments of behemoths such as the United States and China in this area and calls for a European offensive policy. He recommends that AI for industrial use focuses on four sectors: health, transport-mobility, the environment and defense-security.

After in-depth descriptions of the techniques of the field outlined above, the book *Les Big Data à découvert* (meaning Big Data Exposed) (Bouzeghoub and Mosseri 2017) looks into data from different industries that AI will process. These include: earth sciences, ecology, genomics, human health (brain imaging, surgery, oncology, etc.), the individual and society (plant of the future, GAFA, etc.).

This deluge of data, to speak like some authors, requires a deluge of energy. The Internet doubles every three years (Bouzeghoub and Mosseri 2017)! The energy problem becomes crucial. It is estimated that digital exchanges consume about 10% of the energy currently produced, the equivalent of 350 coal-fired power plants of 500 MW each or 180 nuclear reactors (Abiteboul and Peugeot 2017). In addition, the environmental impact of the devices themselves, which are consumers of metals, rare earths and plastics, should also be included. The impact, expressed in terms of greenhouse gases and aquatic and terrestrial toxicity, is little known. These are all challenges that humanity will have to tackle.

5.4. Digital tools and technologies for industrial enterprise

The transition to digital technology and AI has brought in its wake a number of digital tools and methods that have changed the landscape of the company and its means of production. The power of computers, the development of the most diverse algorithms that have made it possible, the proliferation of sensors, and the unprecedented ease of transmitting, storing and processing data have supported globalization, exacerbated the competition and accelerated the transformation of physical and immaterial phenomena. Society and the world of labor suffer changes without any understanding of how far this transition will take us.

Everything can be connected, everything can provide information, everything can be searched for the best... or worse.

Are we entering a civilization of algorithms? What follows cannot cover every aspect of this unfolding revolution. We encourage the reader to consult the following documents which contain plenty of real-life examples (Dal Pont *et al.* 2017a, 2017b, 2017c).

The company requires, in the interest of sustainability, to know the needs of the market, to design, manufacture and distribute innovative products following company social responsibility (CSR) ethics. We shall briefly review some of the most common digital tools behind this management approach.

5.4.1. *Products, innovation, management*

Product Life Management (PLM) techniques (Dal Pont *et al.* 2017b, p. 69) aim to integrate all technical activities of product design and analyze the manufacturing, distribution, use and end-of-life phases.

Innovation is at the very heart of modern business. It could be said that one has to innovate or die.

Crowdsourcing has emerged as a new technique derived from the Internet (Dal Pont *et al.* 2017c, p. 103). It is open innovation: one resorts to external collaboration via the Internet to find a solution to a specific problem. Nothing new under the sun. Louis XIV, who was looking to supply water to the fountains of the Palace of Versailles, put Europe to the test and thus discovered the techniques that led to Marly's Machine which... cost him dearly! This high-profile technique has its counterparts: seekers reveal themselves and the relationship with finders may question intellectual property.

The firm seeks to implement the **collective intelligence** of its employees to find "a source of inspiration". This is **internal innovation**. As such, on June 3, 2019, L'Oréal inaugurated the MYT (Make Your Technology), an incubator where the group employees, already hyperconnected and experiencing digital transformation, can contribute to finding innovative solutions in the industrial sector and in the field of the supply chain (see *L'Usine digitale* from June 4, 2019).

The **supply chain** covers the many processes that range from customer ordering to product delivery (orders for raw materials, supplies, planning and implementation of the equipment needed for manufacturing), as well as related human resources, packaging, transport and handling and tracking, not to mention financial flows and traceability problems!

Just-in-time work, stock minimization, etc. while ensuring customer satisfaction, makes this essential function of the firm even more complex. There are many necessary computer tools that fall under the ERP (Enterprise Resource Planning) umbrella. The concept of **blockchain** has recently emerged: it is a question of securing all block transactions and making all information tamper-proof to fight piracy (see Box 2.2, Volume 1). The management of storage stores (warehousing) is a problem in itself that can be extremely complex from the security and product traceability point of view, as well as that of their handling. The use of robots is intensifying.

Videos
The L'Oréal plant in Karlsruhe: https://youtu.be/iC8TK-edCzo Industry 4.0 at L'Oréal: https://youtu.be/mlDzfKmSYSg

5.4.2. *New tools*

The life of the company in all its forms is now being changed by the advent, not to say by the intrusion, of digital technology, which brings new techniques that we shall review briefly:

– The **Internet of things** (IoT), the industrial Internet: the IoT connects real and virtual objects. The multitude of chemical, physical and biological sensors plays a decisive role both from an industrial and societal point of view. Sensors placed on objects send a variety of information; positioning, physical condition of the object, data about its environment. Combined with Big Data, IoT is a major source of information. A bus company knows in real time where its fleet is. Uber can optimize customer's demands with the nearest available driver. Sensors associated with tracers are capable of sorting plastics (Box 5.4). Box 5.8 gives interesting applications on the use of LiDAR technology in scuba diving. Sensors placed on machines are at the base of M2M (see below). Additive manufacturing enables sensors to be included in the mass of the object and thus can make it "smart". Labeling, one of the guarantees of authentication, is inseparable from distribution: consumers want to know the origin and composition of the product that they buy. Their smartphone allows them to *make the product speak*. Smart Buildings, connected buildings that optimize water and energy consumption, offer people a wide variety of services to improve their quality of life (Box 5.5). Medicine is increasingly using miniaturized sensors; heart rate, blood pressure, blood glucose readings. Google IA Europe believes that deep learning techniques will allow, because of smartphones trained to analyze the images of the eye, to detect cases of diabetic retinopathy (see *L'Express* from December 27, 18). Patients have become "connected objects"; health is now digital. The smartphone stays with the individual everywhere.

– The **dematerialization of information**: the *cloud* at the service of the professional. One of the most noticeable progresses in computer applications has been the gradual disappearance of paper: payrolls, billings, bank statements, directories, tax payments, etc. have been gradually replaced by digital documents. The cloud makes it possible to store and manage a large amount of information about a particular topic with a single click. Box 5.6 shows how the *cloud* can help professionals navigate the jungle of regulations and make the right decisions in the complex area of facility security. The smartphone finds an application in developing countries, in underprivileged regions to collect inaccurate data, to help disadvantaged people in their daily lives (Box 5.7).

– **Virtual reality**: this oxymoron allows a user to be able to interact with and perceive an artificial, imaginary and digitally created world. Very often, users are equipped with a headset to completely immerse themselves in it, or they simply stand in front of a screen. They can move around in a factory, visit Paris, travel down a road, see their home, etc. from a distance. Training is increasingly using virtual reality. Flight simulators that respond to commands have been around for years and enable pilots to be trained more quickly at a lower cost and safely. Kitchen manufacturers, after digitizing an existing kitchen, cooperate with their client to design the kitchen in accordance with their needs and their budget. Engineering produces 3D models (this will be discussed further on). **Virtual prototyping** makes it possible to visualize a part before it is made. It can be modified **at will, its adaptation to its environment can be assessed, and its performance calculated**.

– **Augmented reality**, the **augmented operator**: augmented reality adds virtual elements to a real world, it makes a virtual environment (images, sounds, videos) spatially and temporally coexist with the real world through glasses (head-worn displays), using a tablet (Box 5.9) or a smartphone. The user may be present in the real world or see it from a distance. For example, the operator of a plant during an inspection tour can access all the information about a piece of equipment, whether it concerns its operating conditions, or anything that relates to safety. A corporate device can ask questions and verify its knowledge. This is one aspect of human–machine interaction. The applications are innumerable; a tourist walking through an archaeological ruin can see it as it was in the Middle Ages, or a car driver can see the condition of the road being displayed on their windshield.

5.4.3. *Digital twins*

A digital twin is the digital representation of a complete factory, a piece of equipment, a human being, or even a process. This digital twin represents a process, collects production data in real time, and also monitors performance and sends alerts about drifts.

Videos
"Le jumeau numérique selon Sanofi @ l'Usine extraordinaire": https://www.youtube.com/watch?v=iaAgm4lO3tw

Box 5.10 shows the significance of a computerized model **before** the construction of the factory: it is a tool of choice to come up with the **master plan** of the factory, therefore to represent material flows, movements of people, the ergonomics of workstations, the aspects of maintenance, safety (evacuation plan, firefighting). The same model serves to train staff: for updating information, it will be used to outline the implementation of new facilities. It will be the memory of the dematerialized factory bearing in mind that it was so difficult until then to keep plans on paper up to date.

5.4.4. *Engineering revisited*

In the 1960s, industrial computing eradicated drawing tables, India ink, transparent layers, and physical models reproducing real-scale refineries with the most detailed details. Designers were forced to adapt to the computer and to CAD (computer-aided design) along with its multiplicity of software programs. The advent of the PERT, a task planning tool, is perhaps the most significant event of the period (see above).

Box 5.10 shows how the engineering profession has been profoundly altered by the computer revolution, knowing that **the basics remain the same**. The key points today reside:

– in the **necessity to move quickly** so that the customer is the first on the market;

– in **investments** that are in line with sales volume. In other words, we should avoid investing at the onset in a capacity that will only be reached in a few years... or maybe never!

– in the need for **robust installations** when computer technology by nature tends to make everything complex, fragile and incomprehensible to ordinary people.

Investing may never have presented as much risk as it does in the present day.

Figure 5.5 is an attempt to represent the items that the project owner and prime contractor have to consider in order to be able to properly define the functional specifications.

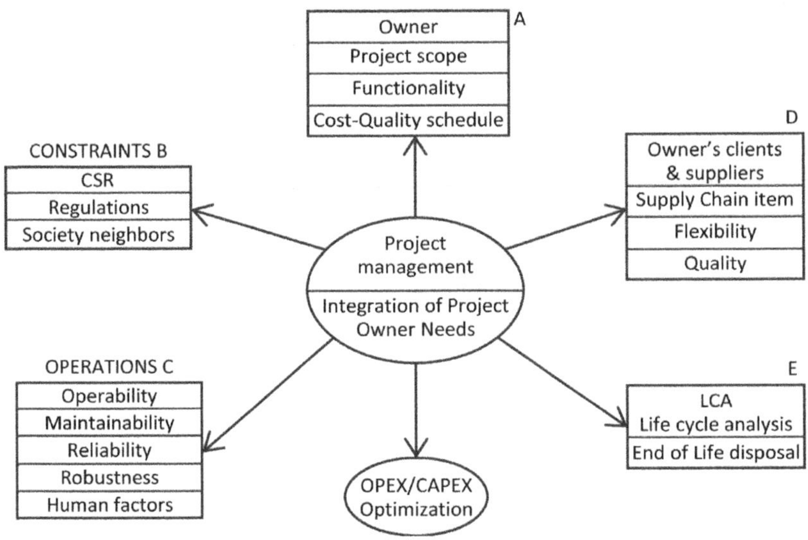

Figure 5.5. *Project owner, prime contractor: functional links. Note: CSR stands for corporate social responsibility*

The **chemical engineer**, a key player in the project in its conceptual phase – the key phase *par excellence* in which all will be decided – must take into account the technological tools that precede it and the operational aspects that we describe in the next section.

The concepts of concurrent or **simultaneous** engineering whose purpose is to reduce completion times, have contributed to writing many books (Prasad 1996). The point is to bring all the actors of the project together from the start and to "crush" the classic phases of the project.

The concept of **fast track engineering** involves the same methods. Experience has shown us that the amount of investment is increased by 10 to 20% because we must consider security factors, we should calculate bigger, and mobilize larger numbers of resources at the same time. Engineering hours are expensive!

Modular construction is preferred by many industrialists. It makes use of prefabricated modules which has the advantage of dissociating construction and assembly, and of building using cheaper or more skilled labor (see Box 3.1). The use of containerized modules, that is, kits on chassis, is similar to playing with Lego. Modules can, in principle, be added as needed and reused at the end of the life of the facilities.

What looks simple on paper actually requires significant industrial experience. The production and maintenance aspects, including the standardization of equipment and automated control are to be considered in parallel.

Building **robustly**, is a necessity. Building a system understood by those who have to make it work and maintain it is also a must. A good system is a system where operators feel comfortable even surrounded by robots, where they know that in the event of a breakdown they can count on qualified people. It is therefore a unified team in which members trust one another to take control of the facility. Up to this point, only human/machine was referred to; nowadays, we have to talk about team/machine.

5.4.5. *3D (three-dimensional) printer or additive manufacturing*

This type of objects construction has spread and become widely popular at a very fast rate in recent years. The reason is still the ever-increasing power of computers.

Originally, stereolithography was invented in Nancy by the team of Professor Jean-Claude André in the 1980s. Today, the object to be built is designed in CAD. The resulting digital object constitutes a file that will control the manufacturing machine. It deposits layer by layer the most diverse materials: plastics, metals, or ceramics (Dal Pont *et al.* 2017a, 2017b, 2017c).

This technique is now used in aeronautics, the automotive industry, the chemical industries (see Chapter 2), jewelry, to manufacture dentures, etc. It has become possible to build large pieces such as bridges or houses. It is interesting to note that the airier the part to be built, the more this technique saves material and that it presents bigger benefits over the techniques for removing material specific to machining. Only additive manufacturing makes it possible to make very complex geometry parts in one piece.

5.4.6. *Robots, robotics, exoskeletons*

A **robot** is a device designed to perform specific tasks, automatically. In "automatically" there is automaton. The Frenchman Jacques Vaucanson (1709–1782) distinguished himself by creating automata, including his famous digesting duck in 1744.

A modern robot is an assembly of mechanical and electromechanical parts powered by energy, controlled by a central unit. It is a mechatronic device combining mechanics, electronics and computers (Wikipedia). From the beginning, robots were designed to relieve humans of repetitive, dirty tasks or those that they were unable to physically perform. The U.S. Army planned in 1969 to use a 3-meter-high walking truck weighing 1,500 kg to transport weapons through the jungles of Vietnam; the driver was directing it with two joysticks (Rid 2016). The project was abandoned at some point.

A robot usually needs sensors to position itself, to capture, to move, and cameras to see as well as actuators (engines, jacks). It can become "smart" by using AI tools.

Robots have invaded automotive assembly lines (painting, welding, mounting), for handling loads, bagging or palletizing operations, and in agriculture (weeding), surgery, or for seabed exploration.

Robotics is the set of techniques that makes it possible to design, build and use robots.

Humanoid robots look like humans. It is not surprising that Japan is ahead in this area, given its technicality and the ageing of its population (Ross 2016). A robot that helps people is sometimes called a **cobot**. Honda's Asimo (Advanced Step in Innovative MObility robot) interprets human emotions, shakes hands and helps the elderly. It can run at 6 km/h. The question arises as to whether one can have an emotional relationship with a cobot.

Figure 5.6. *Asimo (Wikipedia, user: Vanillase)*

The term **cobotics** (or **collaborative robotics**) (Box 5.2) is a contraction of cooperation and robotics: the main object is the interaction between humans and a robot. This type of cobot mainly helps the handling of loads. Widely used for cutting up animals which is a painful operation, it remains dependent on the user.

The so-called **exoskeleton** is an articulated mechanical structure worn by a person and that assists the latter in their operations (Dal Pont *et al.* 2017a). An exoskeleton allows an individual to carry heavy loads or helps paraplegics walk.

5.4.7. *Drones*

A drone is a flying device, or a land, sea or submarine, **unmanned** device, usually remotely controlled from the ground. It can weigh from a few grams to several tons for military applications that we shall not mention.

They allow the monitoring of a territory and the ability to take pictures of hard-to-reach places. They are used in agriculture to monitor crops, or to report the condition of the land in order to know whether to treat and water it. The number of their applications is increasing: rescue operations, and the surveillance of railways, electrical cables and wind turbines (Box 5.3). They will soon be used for the home delivery of parcels.

5.4.8. *Operations management*

5.4.8.1. *Management tools*

Integrated management software packages, commonly known as ERP (Enterprise Resource Planning), the best known of which is SAP from Germany, have enabled the computerized processing of the majority of the usual processes employed in industry: purchases, sales, stock inquiries, production, customers, human resources, accounting, finance. ERPs were born in the 1990s. They are increasingly integrating the cloud, especially for SMEs, for cost reasons.

The arrival of the concept of the "plant of the future" has emphasized the need to control the productive system, globally on the one hand, and at the machine level on the other. Manufacturing Operation Management known as MOM or MES (Manufacturing Execution System) are in use.

The digital twin, already mentioned, makes it possible to move from curative (repairs are done when there is a breakdown) and preventive maintenance (a review is carried out during a halt and after a certain period of operation) to predictive maintenance (the machine warns of the possibility of an outage).

Autonomous trolleys travel in factories carrying raw materials, products being manufactured, or finished products. They use SLAM (Simultaneous Localization and Mapping) technologies (Box 5.2).

5.4.8.2. *Implementing the digital transition*

The suddenness, if not the brutality, of the appearance of the concept "plant of the future", to use French terminology (*usine du futur*), creates a shock wave at the corporate level of the company and even more significantly in its productive system, inherently less malleable, at least from the trade union point of view. The company therefore has to create a **change** with the difficulties highlighted (see Chapter 4).

The book *Voyage dans l'industrie du futur italienne* (meaning Voyage into Italian Future Industry) (Magone and Mazali 2019) describes, based on real-life facts, how the digital phenomenon was assimilated by the manufacturing industry in northern Italy. This industry is the second largest in Europe after Germany. This fascinating journey focuses on the future of Taylorism and Toyotism, the problem of the transfer of human knowledge to machines, the future of workers who find themselves dispossessed and the cultural shock caused by the transition from an oral culture to a writing culture.

It is not possible to summarize this remarkable book in a few lines. We shall only recall a few strong ideas:

– a plant without human workers would be a plant that would stand up on its own, which is absolutely impossible (p. 25);

– a sensory cladding allows workers to work safely among robots. The know-how is in robots (p. 27);

– the key to change relied on a social innovation program focusing on security, training and individual responsibility (p. 29);

– new operators no longer intervene manually, they now monitor (p. 68);

– elders help young people and the young help the elders as long as the latter have experience and the former have agility in digital technology. The experience of the elders has been integrated into industrial processes managed by digital means, thus in algorithms (p. 78).

5.5. Boxes

Autonomous car, human or artificial intelligence

The reliability of the road traffic system

It is a complex system involving infrastructure, vehicles and human intelligence, with cinematic energies involved that translate into road accidents, and whose reliability can be measured in road mortality.

An asymptote seems to appear around 6 deaths per billion km, a level at which France stabilizes, as does the United States: for an average speed of 50 km/h, the risk of a fatal accident is estimated at 30.10^{-8} per hour of driving.

As 1 in 30 deaths is in an accident where one of the vehicles presented a visible technical failure, the order of magnitude of a technical failure with fatal consequence would therefore currently be close to 10^{-8} per hour of operation. The fixed target for car design is 10^{-9} for any failure that involves the lives of road users (Category 4 feared event or ER4 according to ISO 26262 standard).

People or car autonomy

The post-war economic recovery was based on the mobility of people and skills, and thus on the automobile. Today, the mobility of everyday life appears to generate significant costs, but also negative externalities in ecology and road fatalities, hence the dream of a society ensuring the mobility of everyone by means of non-polluting autonomous and connected cars. The search for autonomy for people would lead to making cars autonomous, which raises multiple questions:

– because of its potentially lethal nature, the automobile cannot be considered a connected object;

– the autonomy disappears in the event of an accident, as legal blame will have to be found;

– the concept of polluting emissions is complex and cannot be limited to the local use of such a vehicle.

These three axes must be the focus of further work, but without underestimating their complexity in terms of human factors, of dilution of responsibilities or redirecting local pollution towards other sites or problems.

*ADAS (*Advanced Driving Assistance Systems*)*

These driving aids provide automated driving capabilities in two modes.

The delegation places the driver in a supervisory position: the car takes control, and the driver can take over at any time. This situation is familiar to managers but much less to the average car driver. The challenge is to know how to delegate and regain control wisely, which is all the more delicate since a delegation that goes without problems sometimes results in the driver falling asleep, in being unable to react or regain control in difficult conditions.

For example, the collaborative mode helps drivers stay in the lane by intervening with the steering as soon as the vehicle show signs of crossing a line without flashing a light: less punitive than an alert, this mode intervenes directly on the dynamics of the vehicle without the knowledge of the driver, the latter incurring the risk of misunderstanding! The system must supervise the tactical and operational levels, in a complementary manner to the human brain: if the driver maintains his/her request, the device must know how to abandon the opposition to return to assistance mode without creating an overshoot. This mutual understanding underlies learning, which the casual user of a shared vehicle may come to lack.

Driving automation and dilution of responsibilities

After an accident, insurers and law enforcement often only look for the human factor: the drivers involved should "adapt [their] driving", according to the Highway Code. This accountability of the driver has proven its worth in road safety, but this lever is currently weakening given that driving assistance dilute the responsibilities, thus making its analysis more complex.

The introduction of smart and connected vehicles is a major change in the traffic system and also complexifies the way it is driven: an accident will now have to be considered as a malfunction of the system consisting of infrastructure, vehicles and users.

The automation of driving tasks began almost a century ago with automatic transmissions, but when it comes to replacing the driver of a vehicle which could affect the safety of people in the public domain such as a car in motion, higher demands must be met.

SOTIF (Safety of the Intended Functionality), the new safety improvement lever

As in aeronautics, operational safety has been deployed among car manufacturers and guarantees near-nominal operation closest to the nearest epsilon, imposing the system reliability requirements of ISO 26262, including additional costs related to redundancies and lengthy validations required. This is not enough for consumer automobiles, because the average individual is not professional or well-trained and is therefore unable to regulate what the machine will not be able to handle, such as in an industrial facility or in aeronautics.

The complementary lever provided for vehicles by ISO PAS 21448 is to systematize the confrontation of nominal operation aimed at the diversity of real situations and human behaviors to verify that it will be safe in actual utilization conditions. This SOTIF recommendation is particularly fundamental for the consumer automotive sector, just as important as the safety of operation in this area, although it is still in its infancy for many manufacturers.

Self-adaptability and reliability of the vehicle system

Self-adaptivity, usually achieved by learning, enables a system to adapt to new situations, which may be related to vehicle aging, fuel or improper use. A vehicle, its motor organs or its automatic transmission thus adapt to their environment, adjusting their parametrization, sometimes restricting certain functions or limiting their performance under certain conditions.

The potential for robustness is important with regard to the diversity of real situations and human behaviors: the self-adaptivity of the product to its environment is necessary to meet the requirements of reliability, but this must remain compatible with the diversity of expectations and human reactions, sometimes lost when the system response varies.

Systems that operate differently from the original state, and that sometimes no longer conform to the announced features, raise issues of principle: what to say to customers who have chosen a car for a driver assistance function that compensates for a lack of personal aptitude if the car systematically refuses this delegation out of suspicion about their ability to regain control?

Automated driving and new risky scenarios

It is likely that future driving automatons correctly address frequent accident scenarios, but those where the human brain is able to avoid the accident are by definition poorly known, in the absence of an accident... This kind of scenario is likely to occupy a bigger place in accidentology to come, since the importance of human interactions, implicit communications and visual or audible perceptions is underestimated.

At a time when driving robots are gaining popularity, it is increasingly important to study human behavior when driving, which involves the observation in natural driving, and the construction of massive databases, along with the tools and associated analyses.

No regression will be bearable with regard to the human capabilities learned generation after generation: the development of the Intelligent Transport System must therefore be placed under the control of road safety, which must extend its knowledge to near misses so as to take into account emerging scenarios.

Artificial intelligence and security validations

Artificial intelligence did not wait until the 21st Century to consider the self-adaptivity of gear shifting to conductive behavior, but the issue was accreditation, not safety. One of the advantages of artificial intelligence was already the construction of rules easily parametrizable by a human brain, obtaining in a short time an overall relevant behavior. The current artificial intelligence revolution is not here: it comes from a computational capacity that opens the way to very rapid statistical analyses that provide an opportunity to automate other tasks, especially those related to safety.

Unfortunately, by making behavior dependent on lessons that have been learned, artificial intelligence adds significant complexity, since the vehicle facing a scenario will behave differently from the product sold: the illusion about the safety of the system can be perfect until the moment it proves otherwise in the face of an unexpected situation, when usual interpolation rules between known situations lead, for example, to obvious errors.

Artificial intelligence is both an asset to cover the unexpected situations of SOTIF and a risk that must be constantly monitored regarding security systems, because the certification for an intelligent product on the basis of standardized tests no longer proves the product safety. The construction of parameterized computerized benchmarks that can be used for automatic validation and that best cover the diversity of real situations must be accelerated!

CEESAR

CEESAR (European Centre for Safety Studies and Risk Analysis) is a non-profit association dedicated to research on personal safety, in three areas:

– accidentology, incorporating field collection, statistical analysis, 3D and time reconstruction, accident simulation for evaluating infrastructure improvement or driving assistance;

– human behavior, incorporating vehicle situation analysis, data collection and pre-processing, annotation of anonymized databases and behavioral analysis;

– shock biomechanics, integrating human body tolerances, calibration experimentation for the biofidelity of computer-based models and shock dummies.

CEESAR comprises an independent scientific council and numerous specific certifications covering the activities necessary for this research: use of donations of bodies to science, real-time on-site intervention, access to medical records, processing of personal data.

CEESAR's clients include a historical one (the PSA-Renault Economic Interest Grouping (EIG)) and a variety of other clients who are insurers, equipment manufacturers, public bodies, machine manufacturers, aeronautical designers or other manufacturers in the automotive, motorized two-wheeler or defense fields.

Philippe Chrétien

Graduated as an engineer from the École Centrale of Paris in 1982, in MEDD (Managing Engineering Design and Development) at the Carnegie Bosch Institute in 1998 (Pittsburgh, USA), Philippe Chrétien has been general delegate at CEESAR since November 2015. He is also director of MOV'EO, a member of the steering committee of the DAS Sécurité Usagers Route, chairman of the École Centrale Alumni association in Saint-Germain-en-Laye and secretary of the professional group Centrale Supélec Automobile. He was previously the electrical and electronic systems quality director at PSA.

Box 5.1. *Autonomous car, and human or artificial intelligence (Philippe Chrétien)*

CEA List at the service of the plant of the future

According to a commonly accepted definition, the digital transition of the plant is based on the digital integration of the entire production chain. The objectives are to improve production processes, save time on time-consuming tasks and to secure workstations. Viewed from this perspective, the digital transition is a continuation of industrial automation that has been underway for several decades. However, it also introduces disruption at the strategic level. Because of the digital transition, the plant of the future is able to anticipate changes in demand through a relationship between equipment, operators and customers redesigned in terms of the ecosystem. Whether viewed from the continuity or disruption perspective, the transition to digital technology is a tremendous vehicle for opportunities for the industry.

Digital integration consists more specifically of data integration, also known as digitalization, at all levels of the value chain, from research and development to production and logistics. These three levels will be addressed successively in the rest of the Box. The final section will focus on digital trust, which is essential to the digital transition of the industry.

Digital transformation of research and development

Research and development are the first activity that embraced digital technology in France through the use of digital models generated and exploited using CAD[1] and CAM tools[2]. These models historically made it possible to consider objects exclusively from their mechanical angle and have gradually become more complex to order to integrate thermal or acoustic aspects. In recent years, we have been witnessing a disruptive evolution with the emergence of digital twins. These allow a global and dynamic understanding of objects, but also of the material and processes involved in their manufacturing. All the physical phenomena involved are taken into account through multi-scale simulation, which involves models of different sizes and level of detail. In addition, the digital twin also integrates the time dimension with the ability to simulate the aging of a product according to its use.

The construction of a digital twin is based on the exploitation of data from IIoT[3] sensors that orchestrate the production chain and the product being developed. The manufacturing environment in a wider sense can moreover be reconstructed based on topographical surveys carried out in the factory by drones.

The digital twin introduces a paradigm shift in product design. It is no longer a question of testing successive versions of the finished product, but of having a progressive visualization of its construction in order to carry out at each stage the choices that allow the best dimensioning of the entire production process. Product innovation within the context of the plant of the future is driven by innovation in manufacturing processes themselves.

The digital twin is also revolutionizing the modes of collaboration by allowing teams from several disciplines located in different locations to work simultaneously on the same

1 Computer-aided design.

2 Computer-assisted manufacturing.

3 IIoT refers to the Internet of Industrial Things. IIoT sensors constitute a centrally managed network.

twin. Work sequences are no longer considered successively but in parallel. On the other hand, the digital twin also facilitates open innovation, which is based on collaboration between a wide range of stakeholders, such as large companies and start-ups.

With the digital twin, it is no longer necessary to use physical prototypes, which consequently lowers design costs and reduces the time to market of the new products. In addition, a positive impact is observed on the creativity of designers, to whom no limits are imposed during the design stage.

Digital transformation of production

Optimization of the production line

Apart from the design of new products, the digital twin can be utilized for optimizing the existing production apparatus. The challenge is often to find the best level of synchronization between machines in order to reduce time latency. A project was conducted at the CEA List with the purpose of eliminating bottlenecks observed on an automotive production line. To this end, the engineers implemented a digital twin using the Papyrus platform[4]. This twin allowed the testing of a large number of machine failure scenarios existing in the manufacturing line in order to size buffer zones, thus enabling the debottlenecking of the line.

Reconfiguration of the production line

The plant of the future is changing its production methods from mass series to scalable and personalized products. The digital twin is one of the pillars of this 4.0 flexibility which involves frequently reconfiguring the production tool. All the necessary proofs of concept can now indeed be achieved within extremely short time frames and without immobilizing the production device.

Steering of the production line

In the plant of the future, the running of the automated production lines is managed by industrial control tools such as SCADA (Supervisory Control and Data Acquisition). These are based on cyberphysical systems, which are an evolution of embedded systems[5] integrating additional connectivity and artificial intelligence. SCADA monitoring and control tools are fully autonomous: their digital control is fed by information collected by IIoT sensors to control actuators which then feedback on the line. The deployment of these tools thus makes it possible to automate and make production more agile.

---

4 Papyrus is a modeling software platform (model-driven engineering, MDE) also for the supervision of IoT systems developed by CEA List (video I).

5 An embedded system is an autonomous electronic and computer system dedicated to a specific task, which it most often performs in real time. It has a limited size and its energy consumption is restricted.

Maintenance of the production line

The digital transition is leading to major changes in the field of maintenance. This can be remedial in the first place, namely one could intervene after malfunctions have been observed on a production line. In this case, artificial intelligence provides tools to analyze data collected by IIoT sensors and determine the origin of the outage.

However, apart from corrective maintenance, artificial intelligence paves the way to predictive maintenance. The data issued from the IIoT sensors carry many signs about the degree of obsolescence of the machines and the possible malfunctions to come. The analysis of these data provides a mean to establish optimized maintenance schedules and to free ourselves from routine checks and preventive maintenance operations (change of specific components by prevention) to intervene with machines only when necessary. The anticipation of failures can reach several weeks[6].

Product quality control

Digital integration of the production chain also allows for innovations in the field of quality control carried out on finished products or during the manufacturing process. Online control devices couple optical or radiographic imaging systems with machine learning algorithms which have the ability to conclude that defects in the images may or may not be present using a model previously estimated by means of a limited number of acquisitions. In the context of metallurgy, we have developed a device for the online detection of millimeter rolling defects at the rate of 20 $m.s^{-1}$. The device is based on the analysis of optical images by a network of deep neurons sized with the N2D2 platform[7]. Further control devices rely on the

6 A solution has been developed by the CEA List that can predict the failures of machining machines of the milling type with an anticipation of eight weeks. The project was carried out as part of the Factory Lab, which is an innovation community for the industry of the future located at the Nano-INNOV site, driven by a consortium comprising research organizations and companies (end-users and technology providers). Within the Factory Lab, end-users express concrete and shared needs illustrated by industrial use cases. Research partners (CEA List, Cetim) and technology providers then develop innovative solutions to respond to problems. The Factory Lab website is available at: http://factorylab.fr/.

7 The N2D2 platform was developed by the CEA List. It is the first European open-source deep-learning platform. One of its strengths is that it can facilitate the porting of applications to hardware targets (FPGA, etc.). More details about N2D2 can be found at: http://www.cea.fr/presse/Pages/actualites-communiques/ntic/plateforme-n2d2-deep-learning.aspx. Model-driven engineering information can be found at the CEA List website: http://www-list.cea.fr/technology-research/researchprograms/embedded-systems/design-and-analysis.

comparison of acquired images with images with no defects or, in a more developed manner, with digital models of parts or assemblies[8].

Alternative control techniques are based on data that no longer come from the inspected object itself but are transmitted by the IIoT sensors fitted in the production line. These techniques are capable of detecting and ejecting defective products – or products that are expected to become defective – increasingly earlier, and thus of optimizing the performance of production lines. When manufacturing processes are complex, waste prediction algorithms must be able to handle a massive influx of very diverse and often poor-quality data (some data may be erroneous or even missing)[9].

The augmented operator

The previous paragraphs helped to draw the outlines of a plant of the future made more productive and flexible through the deployment of new digital technologies. Nonetheless, human beings are not excluded from this factory 4.0, in which on the contrary they occupy a central place. If the digital twin is at the service of their creativity, the concept of the augmented operator is intended to increase tenfold their ability to manipulate and perceive.

The first sector concerned by the augmented operator is that of production. In this context, robots intended to provide handling assistance were traditionally intended to replace humans. Today we are witnessing a breakthrough with the deployment of robots that collaborate with the operator in a shared space. The added value resulting from the coupling of human and robot skills is multiplied. The new field of robotics opened by these systems is called collaborative robotics, or cobotics. The help provided to humans by cobots can range from gravity compensation (the cobot supports the weight of the tools) to the amplification of effort. The cobots must preserve all the dexterity of the operator and be versatile in order to be used for different tasks in the factory. The Cobomanip manipulative arm is an example of a cobot. We designed it in partnership with the SME Sarrazin Technologies. It allows operators to move loads of up to 100 kg with submillimeter accuracy. It is used in particular for the installation of components in aircraft engines.

Another example of a versatile cobot is the Hercules exoskeleton, developed in partnership with the company RB3D. This cobot relieves operators when handling heavy loads and assists them in carrying out technical and repetitive movements, such as sweeping movements in the context of public works. In general, cobots contribute in making factories

8 Such a control method can be implemented, in particular, by means of the SIRIS solution proposed by the CEA List start-up Diota and Actemium: https://www.actemium.fr/fr/toulouse-robotique-automation/siris-2/.

9 See the international challenge proposed by Bosch and organized by the Kaggle web platform: https://www.kaggle.com/c/bosch-production-line-performance. The CEA List ranked in the top 5% among the more than 1,000 teams that participated in the challenge.

safer places, thus reducing the health risks of musculoskeletal disorders (MSDs), and in making them more open, by expanding the range of potential operators (workers with disabilities, seniors).

Beyond physical assistance, the augmented operator benefits from a cognitive assistance provided by augmented reality[10], which consists of superimposing virtual information on a real object. Equipped with this technology, for example, an operator with an adequate vision system can be guided through the successive stages of manufacturing a product[11]. The solutions proposed by the start-up Diota, based on tracking technology that was developed at the CEA List, allow the operator to visualize drilling locations or to be informed of the misplacement of a part. In addition to augmented reality, cognitive assistance can be done through proactive voice assistants who can interact with the operator. The efficiency of these systems depends on their ability to analyze the work environment to deliver relevant information at the required time.

Finally, assistance to production requires new training methods made possible by virtual reality. This allows the immersion and evolution of an operator in a totally virtual environment whilst replicating all the components of the real world. Operators interact with digital models through tools such as the Mandarin force-feedback glove that we have developed. The haptic feedback can be capable of mimicking the touch of different textures. Based on virtual reality, the training of operators is done without interruption of production. Moreover, virtual reality can also prepare people for dangerous operations[12] and enables the anticipation of problems that may be encountered.

The augmented operator also finds applications in the field of maintenance. A mobile medium such as a tablet or smartphone can be placed by an operator in front of a device to be controlled in order to enable the visualization on the screen of the manipulation sequences to be performed. 3D goggles can also be used, with the advantage of leaving hands free to the operator. One can also interact vocally with the device, for example to validate control points.

Decision support

Factory 4.0 sees the development of expert systems designed to provide decision support to operators. To establish their recommendations, these systems mobilize a set of rules in which knowledge from human expertise, inspection reports, standards, etc. is compiled. Expert systems are particularly found in the maintenance field, especially when it proves costly and dangerous. At the CEA List, we have developed a decision-making tool for a gas

10 See “Augmented reality in 90s” by the INRIA: video II.

11 For this topic, see the solution “Digital-Assisted Operations” developed by Diota.

12 In a different context than that of a plant, a tool developed by the CEA List should be mentioned because it provides a form of training to drop equipment at the bottom of the sea by means of a crane placed on a boat undergoing rolling.

transmission company with the purpose of providing recommendations about the pipes to be excavated and to be replaced due to corrosion risks. This tool was designed using the ExpressIF platform, dedicated to the design of decision support software making use of fuzzy logic[13]. The potential applications of expert systems in the plant of the future fall far beyond the field of maintenance. They can also help sizing production lines, allocating resources or even sizing tasks.

Navigation in the plant

The lines of the plant of the future are increasingly taking the form of reconfigurable islands dedicated to assembly, machining or control tasks. Navigation between these islands is provided by autonomous trolleys with flexible trajectories. Autonomous navigation is made possible through cartography and simultaneous localization SLAM technologies (Simultaneous Localization and Mapping), which allow machines to build a map of their environment and at the same time to locate themselves within this environment without necessarily using GPS-type information.

Digital transformation of logistics

The previously mentioned SLAM technologies also have applications in the field of logistics, particularly by enabling the delivery of products between factories and warehouses. The digital transformation of logistics also requires the establishment of digital continuity between the tools governing every aspect of the plant ecosystem. ERP software, which manages orders, purchases, logistics and can also integrate commercial aspects, thus communicates directly with WMS systems, responsible for inventory management, and ESM systems, in charge of production.

Digital trust

Digital transformation is leading to an increasing connection of production sites to the outside world, resulting in increased exposure to cyberattacks. When they do occur, they spread very quickly to the entire plant due to the interweaving of the different information and supervisory systems, such as the resource management system (ERP), the production monitoring system (MES) and the machine control system (SCADA). Potential consequences can range from the temporary blocking of a production line to industrial sabotage. The historical cybersecurity solutions of computer systems are unsuitable for the specifics of industrial control systems and need to be completely redesigned.

13 Fuzzy logic contrasting with binary logic can be seen as being closer to the modalities of human reasoning.

The remedy consists in an industry 4.0 secured by design by means of tools that can mathematically guarantee the quality and reliability of software, such as the Frama-C[14] platform developed at the CEA List. This has been the subject of a positive evaluation by NIST and has been used in particular in the fields of avionics (Dassault Aviation) and nuclear (EDF).

Videos

Papyrus video presentation:

http://www-list.cea.fr/systemes-embarques/195-presentation-de-papyrus

"Augmented reality in 90s" by the INRIA:

https://www.youtube.com/watch?v=vmaEI-06C0Y

CEA List

The List is one of three technology research institutes of CEA Tech, the CEA technology research directorate. Dedicated to intelligent digital systems, its mission is to achieve technological developments of excellence on behalf of industrial partners, in order to create value. Its 820 employees accompany 200 French and foreign companies each year in applied research projects in five fields: artificial intelligence, the plant of the future, innovative instrumentation, cyberphysical systems and digital health. As the Carnot Institute, it invests 25% of its budget each year in scientific resourcing actions, to identify the disruptive technologies of tomorrow.

Mounir Bakkali

Mounir Bakkali holds a degree in Electronics and Computer Engineering from the CNAM (French National Conservatory of Arts and Crafts). He is also an atomic energy engineer at the French National Institute of Nuclear Science and Technology. He began his career in the

14 See the presentation of the Frama-C platform at the CEA List website: http://www-list.cea.fr/systemes-embarques/196-presentation-de-frama-c.

semiconductor industry working for three years as a photolithography process engineer at Altis Semiconductor. He joined the CEA seven years ago as a Research and Development Project Manager. He is currently Head of Industrial Partnerships at the CEA List. He is the co-inventor of six patents and co-author of 10 scientific publications.

mounir.bakkali@cea.fr

Box 5.2. *CEA List at the service of the plant of the future (Mounir Bakkali)*

General Electric Digital: industry 4.0 approach

"Industry 4.0" is the term created by the German government to characterize and promote the Fourth Industrial Revolution. The aim is to interconnect the plants, collect data and capitalize on cloud technology to raise the level of automation. Gradually, the concept becomes wider to include other areas of industrial production and software used to improve operational performance, increase revenue and reduce production costs.

The digital industrial transformation describes the continuous evolution that brings the company to level 4.0. This evolution is technological and human and affects all sectors of the company.

This evolution can be described based on four technological themes:

– machine connectivity in the factory;

– improved production processes;

– equipment maintenance optimization;

– the augmented operator;

and four organizational themes (which are not discussed here):

– the CEO's involvement;

– the change of managerial approach;

– the development of software DNA within the company;

– the improvement of the employees' skills through continuing education.

Machine connectivity

This is the founding element of the transition towards the digital factory.

The principle is to have an exhaustive view of the condition of the machines and of the state of the processes in the factory. The HMI/SCADA software programs consolidate all this information and monitor the condition of the machines. They help the operator maintain the systems as operational.

The United States estimates $20 billion per year of losses due to unannounced shutdowns in its production cycles. About 80% of these shutdowns can be avoided, especially since 40% of them are the result of human error. The latest generations of automation software control the full spectrum of machine parameters and are increasingly operator-centered. The operators identify problems and their causes more quickly, and they benefit from on-demand support. Software programs provide "smart" alerts and limit the potential for human error. The analysis of operational data is a means to identify problems preventively.

Connectivity and automation software make it possible for a company to become involved in the digital transition. The use of cloud technology enables all the data to be accessed and the operational statuses of connected factories to be visualized. The next step consists of improving production processes.

The management and steering of production operations

The improvement of production processes is usually achieved with software for managing production operations (Manufacturing Operation Management – MOM or Manufacturing Execution System – MES). These software programs collect and analyze production data in real time to transmit key production indicators to the operator. These indicators are mainly related to process and product quality, machine efficiency, product traceability and batch analysis.

Once again, the latest generation of production management software is designed around the needs of each operator function. The interface is customized. When the operator arrives at his/her position, his/her production plan is already displayed. The machine is started, production is initiated. The software assists the operator on production and displays the current performance, the planned one and the history. Depending on the values of the indicators, certain tasks are to be done. The software guides the operators to help them return to optimal production levels.

More and more processes and machine data are being collected in the cloud. On the one hand, this allows getting an overview of the production processes or visualizing what is happening in the different factories. On the other hand, it makes it possible to compare the processes of one production line to another one, or even to compare the performance of equivalent plants. Finally, the analysis of the collected data helps identify the characteristics of optimal processes. Solutions can be deployed on the least performing sites or channels.

Machine data collection is also at the basis of the design of the digital twin: it is the digital representation of a real object or system. The concept is used for simulations/analysis of hypotheses (what if) and predictive maintenance. The digital twin is an additional tool to improve productivity. It optimizes the maintenance of the machines.

Equipment maintenance optimization

In general, manufacturers apply the maintenance recommendations of their machine suppliers. With experience, they realize that their maintenance costs are overstated. However, they do not know what steps to take to reduce these costs because they do not have a global approach of the maintenance of their machine. They are not able to assess the implications of a change in procedure.

Asset performance management software is available. They help the operator find the best compromise between reducing maintenance costs, the risk inherent in equipment and its production. As a matter of fact, maintenance ensures the performance of the equipment. However, this maintenance comes at a cost. There is a trade-off between maintenance and production. A third component should be taken into account: that of the risk taken by not sufficiently maintaining the machine which could lead to an involuntary interruption.

The best ones meet ISO standards for continuous process improvement. This process is based on four pillars:

– the identification of equipment criticality;

– maintenance strategies to be put in place;

– operational execution;

– feedback information analysis.

In identifying equipment criticality, every piece of equipment ought to be ranked in order of criticality and a centralized view of each one should be available. Some machines are more critical than others. An involuntary interruption or a failure of some machines can have a serious impact on production and/or the environment. Other machines may fail but do not jeopardize production. This classification allows one to identify the maintenance strategies to be put in place.

There are maintenance policy libraries that exist which cover a very comprehensive number of equipment. These libraries are made to be integrated into the maintenance optimization software. Strategies range from the simple monitoring tower including meter readings to be done to predictive maintenance, not forgetting conditional maintenance.

Predictive maintenance resorts to digital twin technology. This twin makes use of a very detailed model of the equipment (blueprint). The model takes into account the type of materials that the machine is made of, operating conditions (temperature, pressure, humidity) and further components specific to the machine. Based on historical data, the model determines the aging curve of the actual equipment. The digital twin permanently monitors the actual equipment and alerts the operator of future failures early enough for an intervention to be made as part of a scheduled maintenance cycle.

By implementing the different maintenance strategies to be put in place, the software can identify redundant actions and eliminate some sometimes very expensive interventions.

The technology allows maintenance operators to be connected to the system supervision center by mobile tools (phone, tablet, or augmented reality glasses). Their interventions are optimized. The software leads them directly to the location to be controlled via geolocation technology. In complex factories, this support helps to save time. It also ensures that the operator is constantly in a protected area. Once in place, operators who find themselves in difficulty can consult a tutorial on their tablet or their augmented reality glasses. Online support is also available.

Every action and comment made by maintenance operators are listed in the software. Equipment criticality as well as the strategies put in place are re-evaluated according to practical experience. This ensures continuous improvement in the maintenance process.

Videos

Additive printing

3D printing technology, also described as additive, involves superposing layers of materials rather than removing some from a block of material. This allows complicated design to be produced more efficiently and with better mechanical properties. With the traditional method, it sometimes takes several parts to make a more complex one. Now, 3D printing makes it possible to produce this complex part in one piece. In the future, the 3D printing process will become faster and the properties of the materials used will continue to improve. Previously unimaginable pieces will be created.

Visit the GE Power additive manufacturing/3D printing plant in South Carolina and learn about the production process of a part for a gas turbine starting from its digital design. 3D printing processes are still quite slow. Faster production with better materials will be possible with this technology in the future.

https://www.youtube.com/watch?v=pBmLHDANDKM

M2M w/GE Predix Platform

The GE Digital Predix platform aims to connect individuals, machines and data.

Industrial platforms, such as the one proposed by GE (Predix), enable machines and systems to be connected and applications to be developed in order to create new services and optimize existing ones.

https://youtu.be/dIydQjHmyXo

Asset Performance Management

The GE Digital (Predix APM) equipment management software is connected to the machine and retrieves its data. The analysis results in: 1) determining the overall condition of the machine: 2) assisting the operator with predictive maintenance (or more specifically, to inform the upstream operator of potential damage so that he/she can implement corrective actions); and 3) assisting the operator with a series of maintenance optimization strategies.

https://youtu.be/LnVZaoiClUU

MES

Visit the GE Aviation plant in Cheltenham in the UK and learn how the GE Digital industrial execution systemmakes it possible to work smarter, more efficiently and improve productivity.

https://www.youtube.com/watch?v=tkq56lE0Nb0

Digital twin

Electricity networks are becoming more and more complex and operate a mix of different energy sources: gas, nuclear, renewable, etc. The different sources are traditionally managed *in silo*. In order to meet production and cost reduction needs, and to better manage risks, energy suppliers need to have an overall view of their network. The networking digital twins of GE Digital consist of software programs that connect the different sources of the physical network to the digital network. The digital twin accumulates over time the operational data of the equipment as well as the environments in which they operate. Data analysis and the continued use of machine learning technology enable operators to be supplied with key information and operations to be optimized.

https://www.youtube.com/watch?v=5Ji8dZqkECM

Here is the demonstration of a digital twin from GE Digital for a remotely piloted offshore wind turbine. The wind turbine uses seven different engines. For each one, a digital model has been created. The model uses technical and physical information. The simulation shows how the performance of an engine deteriorates over time with the increase in coil temperature. The digital twin provides means to warn the operator about the loss of engine performance, to anticipate its end of life, by constantly measuring the temperature of the engine coil. The operator can take action well ahead of when time is critical.

https://www.youtube.com/watch?v=P36yJkE1zlM

IoT

Connected elements are being developed in every sector. Data analysis reveals previously unsuspected information. Many applications are emerging to deliver new services.

https://www.youtube.com/watch?v=ghwkQdtG85Y

Drone & Tech

Drones are now part of the array of inspection tools of the industry. They provide access to hard-to-reach or dangerous places and maximize the inspection of industrial sites. One should consider how drones can efficiently and safely inspect a site of 60 wind turbines in Western Canada and how GE Digital's preventative maintenance provides energy distribution to more than 54,000 homes.

https://www.youtube.com/watch?v=6JqUTpp0t5A

GE Digital

GE Digital is reinventing the way manufacturers build, operate and maintain their equipment, by capitalizing on machine data to reveal relevant information that can achieve key results for the business. GE Digital's Predix portfolio – which includes the optimization market flagship applications for equipment maintenance and field service management – enables its customers to optimally manage the entire lifecycle of their equipment. Strengthened by Predix, the application development platform designed for industry Internet, GE Digital enables business industries to operate faster, smarter and more efficiently. For more information, visit www.ge.com/digital.

Emmanuel Coury

Passionate about the digital transformation of the industry, Emmanuel Coury is the Head of Sales Operations and Strategy at GE Digital Europe.

www.linkedin.com/in/emmanuelcoury

Box 5.3. *General Electric Digital: industry 4.0 approach (Emmanuel Coury)*

Tracers for the plant of the future and the circular economy

Why tracers?

The economy is pushing for producing components or products that are increasingly efficient, lighter and ever more reliable, with demonstrable traceability and time savings, without the use of sophisticated means (without sensor settings and data acquisition means), without destroying the object, without lengthy inspections and preferably controllable at any location.

The main, or more pressing, requests concern:

– authentication: the fight against counterfeiting (intellectual property);

– traceability: the responsibility of the producer, responsible for himself/herself but also for his/her subcontractors;

– the product recycling or end-of-life in response to the need to enrich the circular economy.

There are existing technologies to convey information about industrial products such as RFID chips or barcodes, but they intervene at the end of the product lifecycle or require laboratory treatment (DNA). Therefore, one perceives the need or the interest in using tracer-based technology that would then become the center of a global strategy.

Principle and definitions: marking-tracing procedure

The **need**: a strategic tool that allows a product, a material, a component to be traced at a chosen stage of its development; it then becomes possible to identify it formally to the demand – eventually in later lives, preferably without resorting to lengthy and costly analyses.

The **principle**: it is the introduction of a marker-tracer, or a mixture of tracers. The latter are specific chemical compounds or a composition of tracers that, when excited by a UV or infrared source (transmitter), it re-transmits radiation – usually in the visible spectrum that is analyzed by a dedicated reader (detector or sensor). Some tracers may be excited by radiation in UV and re-emitted into the visible spectrum (Stokes), or others may be excited by wavelengths in the infrared range and re-emitted into the visible gamut (anti-Stokes).

It is preferable to choose tracers with characteristic signals that can therefore be easily identified or differentiated by dedicated detectors. Chemical inertia and thermal stability (700 to 1,500°C, for example) are also key parameters.

In addition, it is possible to produce combinations of proprietary tracers in large numbers, namely unfalsifiable or impossible to copy, with well-defined spectral characteristics (excitation domain, peak emission width of the order of the nm, multiple peaks or not, re-emission time). In many cases, only the person who introduced the tracer can find it. Without this particularity, the number of combinations would be very limited and the tracers subject to copy.

The sources of excitation and detectors can be fixed and the object (the particle) ejected by air jet (in the case of sorting lines) or portable and integrated with auditory validation or a light diode (anti-counterfeiting strategy).

Strategy and selection: ecodesign

It should be noted that tracer-based marking is highly strategic, even if it sometimes only occurs in the final stage of product completion.

Surface marking

Occurring during the final stages of product preparation (or packaging), it is usually employed during the labelling or parison stage. The tracer(s) can then be integrated into a printing ink or one of the components of the product.

Mass marking

Mass marking has a serious advantage: only industrial or controlled production makes it possible to homogeneously dose one or more tracers, hence the difficulty of counterfeiting. However, the amount of tracer required is obviously higher, and therefore so is the cost.

Compatibility of different simultaneous markings

Mass marking does not exclude one or several others on the surface. As such a material originating from a producer can proceed into several different industries that will be separately identified.

A product composed of several components can therefore see the chain of one or more of them perfectly traced.

Applications

Tracers allow the coexistence of multiple markings (material origin, transformer(s) identity in the chain, destination at the end of the lifecycle, etc.). Label marking containing specific markers is of course not only possible but also complementary.

Plastics

There are numerous applications of tracers in plastics which have been demonstrated in pre-industrial and industrial stages.

For recycling purposes, clear PET bottles, PVC, PE and multilayers could be traced in the mass with combinations of tracers of a few ppm, without modification of color or mechanical properties and then sorted in the form of pellets at industrial sites.

Finally, a demonstration was made of the fullness of this method, by carrying out black sorting. This consisted of the high-speed demonstration of sorting particles (2.5 cm of average diameter) of black polypropylene created by the automotive industry (TRIPTIC program with Tracing Technologies®, Renault, Plastic Omnium, ENSAM, etc.) in conditions representative of the manufacturer, that is at a scrolling speed of 3.5 meters per second and at least an efficiency of 95%.

Packaging

The possibility of using a marking strategy allows one to both mark the packaging for recycling purposes (type of monomaterial, multilayer, content, contaminant, recycling process) and its control during its later lives (reuse, recovery, etc.). It is a means to meet in particular the requirements of the European Directive 94/62/EC on packaging, which aims to limit environmental risks.

Recycling: new materials

Regulations impose recycling and recovery targets. This would thus seem to limit the introduction of new, more efficient products because the creation of a new sector cannot be established without a control of the flows. The use of new combinations of tracers makes it possible to choose a process at a certain point (low flow for example), and then divert it to another more favorable process at that moment. There is positive sorting and negative sorting, but also prospective sorting.

Such a product intended at a given point for disposal can then become a recognized secondary material and therefore recovered. There are then no restrictions on creativity. In short, the use of tracers contributes to a positive evolution of creativity and strategic reorientations.

Tracers and 3D printers

3D printing is now part of our daily universe. Printers are capable of using consumables in the form of filaments and powders. They mainly use thermoplastics whose melting point is about 200°C, but also new composite filaments based on carbon or composite fibers. Finally, they comprise a UV laser.

In summary, it is a preferable environment for these selected tracers in order to be stable at high temperatures and to UV-visible radiation and whose granulometry is compatible with the nozzles being used.

This should boost the fight against anti-counterfeiting by integrating tracers directly into the initial material or by producing on-demand mixtures.

Health: safety

Due to both chemical (no migration, insolubility) and thermal stability (area well above 700°C, usually above 1,500°C), tracers are generally suitable to materials for food contact. Some have already received approval.

Tracers and society

Health, safety and the environment require evidence about the effectiveness of the tools used. In the face of exchanges travelling all over the planet, reliable, controllable tools are needed.

Sensors must be either easy to use (use, counterfeit or control) or fast (sorting lines).

They respond to the demand for trade globalization and environment protection.

Conclusion

Consider the balance of the mainly Stokes and anti-Stokes tracers:

– both thermal and chemical stability, hence a strategic and adaptable tool: the fight against counterfeiting, recycling or disposal;

– industrial tool (high speed, automaticity on finished product as on crushed material);

– possibility of multi-level readings.

Tracers thus constitute an indispensable tool in the technology toolbox.

For further information

Maris, E. (2014). Écoconception par une approche multicritère d'un système polymère/traceurs/détecteur pour le tri automatisé par spectrométrie de fluorescence UV-VIS en vue d'augmenter la recyclabilité des produits. PhD thesis, ParisTech, ENSAM.

RDC Environment (2017). Material efficiency by marking in EU Ecodesign. Report, Dutch Ministry for Infrastructure and Environment, The Hague.

Techniques de l'Ingénieur (2011). Traçabilité des produits industriels par marquage chimique [Online]. Available at: https://www.techniques-ingenieur.fr/actualite/articles/tracabilite-des-produits-industriels-par-marquage-chimique-25806/.

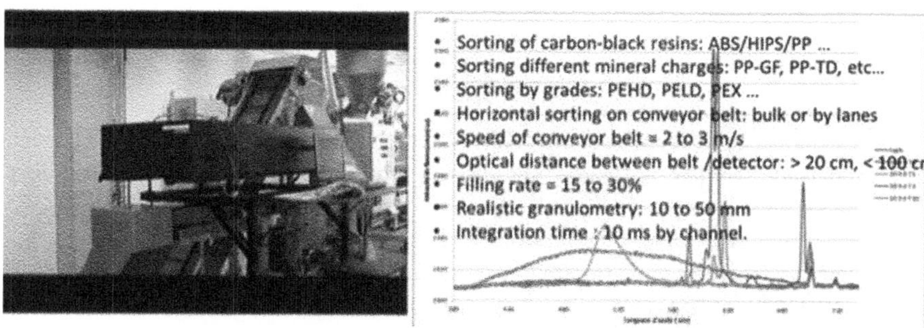

Figure 5.7. *Automatic sorting of black plastics (source: Tracing Technologies®). For a color version of this figure, see www.iste.co.uk/dalpont/process2.zip*

Figure 5.8. *Fluorescence under UV excitation. For a color version of this figure, see www.iste.co.uk/dalpont/process2.zip*

Tracing Technologies® (TT)

Tracing Technologies® (TT) has developed positive, instantaneous and contactless authentication and identification processes for products and materials. These innovative processes break with conventional fluorescence-based marking and reading approaches. They provide an unequivocal answer to the unique registration of a product. These are invisible marking processes, in the mass (except metals) and/or on the surface, by markers in extremely small quantities (a few ppm of concentration).

www.tracingtechnologies.fr

Claude Lambert

Physics engineer and chemist, former R&D Executive for Alsthom, Exxon and Occidental Petroleum, Claude Lambert is Head of the Plastics Division and Environmental Coordinator at the French LNE.

He is also an expert with ties to national and international organizations: ANVAR, ADEME, ONUDI, AFAQ, AFNOR. A former French representative of the European Packaging Standards Committee and Chairman of the European Group of Heavyweights and Dangerous Substances, he is a scientific expert in EU research and development programs as well as an expert with the UN-WTO on the topic of packaging issues.

Finally, he is co-inventor of five international patents for markers/tracers and co-founder of Tracing Technologies® with Jean-Michel Hachin.

Box 5.4. *Tracers for the plant of the future and the circular economy (Claude Lambert)*

Smart, did you say smart?

Or, how will the Smart Building be the beating heart of the Smart City?

The situation in the world imposes an obligation on us to make a difference.

This is no longer really a matter up for debate, today everyone agrees that our civilization issued from the Industrial Revolution has accomplished its full potential and is now at the boundaries of what is tolerable. The megatrends of the time are well-known, first and foremost with technology, which since the beginning of the 21st Century has been generating a new breakthrough, one that has even become known as the Fourth Industrial Revolution. The challenges to be addressed for this industry 4.0 are indeed enormous!

When it comes to the demographic aspects, the global phenomenon of metropolization is beginning and accelerating, bringing along with it increasingly complex problems, which only the emergence of artificial intelligence and digital technology will be able to correct.

Although obviously different in their quantitative and even qualitative aspects depending on whether we observe them for developing or so-called developed countries, we find all around the globe, our common home, the issues of displacement, traffic, transport times, access to water and energy.

While this is not completely new and it is important to leave behind the romantic idea that hypothetically "it was better before", the big difference is that with access to instant information, today everything gets to be known and very quickly.

It is not absolutely certain that the air breathed in the Paris of the Middle Ages was so much more pleasant to breathe than that of today, but the inhabitants of the time did not have access to an application giving them live the level of fine particles or ozone particles.

China is often cited as an example of a country where things are "wrong", but sometimes we forget to mention everything that is innovative therefrom, and one should not abstain from thinking that from the problem will emerge the solution. Chongqing, with its 34 million inhabitants, or almost half of the population of France, in an area equivalent to Austria, aims to become a garden city by redistributing 20 million of its inhabitants over small urban areas in the countryside. In Liuzhou, a "green city" designed by Italian architect Stefano Boeri, will be built in the forest and 40,000 trees will be planted to reduce the pollution cloud. Based on China's growing mastery of technology, there will soon be dozens of new "smart" and virtuous cities.

On a smaller scale, in the Ile-de-France (Greater Paris) region, with its 12 million inhabitants (making it one of the first European agglomerations), the situation is bitter in terms of health and well-being. As a matter of fact, there are 48,000 premature deaths per year due to poor air quality. This probably explains the wishes expressed by many Franciliens[15] who want to leave their region!

On a more global scale, at the environmental level, it can already be noted that:

– 50% of the city area is dedicated to the individual car;

– 10% of the population is already known to be electro-hypersensitive and the rate could reach 50% by 2050;

15 A Francilien is the name given to a person who lives in Greater Paris.

– 30% of the population suffers from allergies due to pollution; at this rate it would be 50% in 2050!

From an economic point of view, the impacts are not negligible:

– a significant portion of the population says they have problems or fears of finding themselves in a precarious situation;

– more specifically, for Franciliens, 33% of their income is on average spent on housing and they spend more than 1 hour 30 on transportation per day, with more than 60 hours per year spent in traffic congestion (these figures are sometimes worse for other global metropolises). This generates an increasing rate of absenteeism (4.5% in the private sector and 9% for local authorities) which costs businesses 45 billion euros.

We therefore urgently need to rethink the city and create a more desirable urban space, which puts people and sustainable development at the heart of project design.

Technological change is taking place everywhere, and the digital transformation of society is irreversible. The only question that remains is, when will it be possible to observe major transformations in our worlds:

– Will clean, autonomous vehicles be on our streets in 5 or 10 years? It does not really matter; in fact, the real answer lies in the certainty that they will soon replace the noisy and polluting cars of today.

– How many self-employed/independent/multi-active workers will there be in three to five years? It is understandable that paid employment for life has relatively little future.

– When will real-time information on the environmental parameters of the city and buildings (consumption, production, etc.) and the emergence of induced services be available? Three to five years? The first Smart Grids are already operating and the first Smart Cities such as Dijon have been inaugurated.

The new so-called digital native generations, and with them the advent of the era of start-ups, are pushing for rapid change and the emergence of new uses and services. This globalized craze for new technologies generates a scale effect that will contribute to the acceleration of this transition towards what can already be conceptualized as 3D territories (digital, decarbonized, and decentralized). Along with them, other questions arise:

– Decentralized energy production: in three to five years?

– The generalization of blockchains as a new transaction mode: in three to five years?

With all these ongoing transformations, we realize that the building is also subject to the laws of the Internet, and this changes the game and the habits of brick and mortar. Now that they are becoming smart, buildings will be the beating heart of Smart City.

It is however necessary to propose a definition for the Smart Building, if we want to avoid making it simply a trend, which, as everyone knows, never last very long.

A building can be described as smart or intelligent, regardless of its destination or its primary/main use (housing, workspace, living space, production space, etc.) that follows the concept of sustainable development throughout its lifecycle, from construction to dismantling, and which integrates digital technology into its operation in order to optimize energy consumption and interaction with its environment.

In short, it is a building of the digital generation that thus integrates the Internet as the fourth fluid.

We move from "water and gas on every floor" to the concept of real-time and secure steering by means of the Internet protocol.

As such, the building is ready to integrate a whole series of connected objects and thus to deliver real-time information (data) and services, this is then referred to as Smart Building Ready2services.

The times has come when the actors have finally understood that the strategy of everyone for himself/herself with one's proprietary system, closed and not communicating with the products of the neighbor, is vain and a dead end. Time has come for pooling and sharing data between all stakeholders, from design to operation, to execution and obviously not forgetting construction (BIM – Building Information Modeling). This has already happened with information technology, where no one no longer wonders if a computer of such a Korean manufacturer will be able to print a document on a Chinese printer using American software. Regardless of the brand of my computer or my smartphone and my internet service provider, I can listen to the music I want, download the app that I want and send e-mails to whoever I want. This is the advent of interoperability. At the building level, this requires manufacturers from the relevant different sectors to accept the same concept and open their "chakras" forgetting the old "every man for himself" reflexes. In the end, it does not matter what the protocols and the technology are, and what matters to the end user is that the different products in the building or apartment communicate with each other. It is the entire lifecycle of the building that is being affected in its approach, as is the case of contemporary society where everything speeds up faster. Its usage cycles are becoming shorter and the building must be flexible, modular and scalable. By logging in, the building becomes smart. There has been a shift from home automation and other automatons, too expensive and too complicated for everyone and quickly obsolete, to the Smart Home and Smart Building world for all.

The success of the digital revolution will make it possible to make housing smarter.

Analogously to other industries that are doing the same, or for some that have already adopted it, the building is forced to adapt to the rules of the Internet game to transform itself and become:

– a multi-purpose Ready2services service platform: from home support to anti-intrusion security management including preventive or predictive maintenance and air quality information, every installed system will need to be open, reversible, interoperable and mutualized to limit the extra costs of installation and operation;

– connected and communicative while ensuring the cybersecurity of its big data;

– oriented towards its users in a simple and user-friendly way.

At the Smart Home level, in this way, each resident will be able to customize their home and add the services they will need or want, at marginal cost, by upgrading their basic facility with IoTs from various sources. The recent advent of Google and Amazon voice assistants that precede the offerings of other giants in the sector demonstrates that this story is just beginning and will impact the world of Smart Home. All these connected objects dedicated to housing will, in addition to their basic function, enable us to generate new services for the inhabitants, with a main focus on comfort, health, well-being and safety:

– security: water leak, smoke, gas, intrusion detection, connected lock, etc.;

– health and well-being: air quality, humidity, e-health, etc.;

– comfort: lighting management, windows, heating control, energy expenditure forecast, etc.;

– assistance to persons: fall detection, light path, health monitoring, etc.

Every era has its dreams and its tools. With the advent of Industry 4.0 and its emerging technologies, the Smart Building will generate new uses.

Analogously to the automotive industry, which is moving towards a mobility-centred service economy (who is the suburban who still wants to buy a car?), the construction world is moving towards a service industry by offering plural, scalable, reversible and multi-purpose spaces.

The generalization of teleworking as a response to the problems of mobility in large urban areas and the improvement of the quality of life imposes a revolution in the design of public housing. Even if one is nomadic, one has to plan where the PC/office is going to be installed. Paradoxically, these few additional square meters will make a deep impact on the residential real estate development industry, which must find a response to their financing issues!

The building will become a support for services around its various spaces. In addition to becoming energy-efficient (we are talking about buildings with positive energy, that is, they produce more energy than they consume), the concept of a building with a positive economy, which allows the generation of income other than renting by the square meter alone, can thus emerge.

The condominium will be able to really take advantage of the surface areas of its unused roof terraces to install photovoltaic panels and make a profit from them, either by using the energy produced in the form of self-consumption or by reselling it by injecting it into the grid. Similarly, by employing real-time data, the building can offer temporary parking spaces not in use by its residents, charging services for electric vehicles, terraced vegetable garden areas and potentially unused common premises.

Everyone can easily picture from watching the emergence of start-ups that new trades will *emerge to provide these services and put into practice the concept of* Building as a Service (BaaS).

The connected and communicating building is no longer a superficial luxury item reserved to an elite, it has *become a reality with several thousand homes delivered and inhabited. This becomes essential for many inhabitants and in fact constitutes a real value proposition of some real estate developers.*

Independently of its original vocation (tertiary, services, homes, shops, etc.), within three years it can be reasonably announced that any new building built in France will, in one way or another, be connected and intelligent. Just as cars are connected and become plug and play communication platforms through their users' smartphones, for their part buildings will also benefit from being communicative in the same way. How can it be explained that it is possible to stay connected to the Internet effortlessly through Wi-Fi on the TGV (French high-speed train) or in a moving Airbus, but that it would remain complicated to get an efficient connection in one's new home when moving in? Everyone (owner and/or user) will expect a building to be connected as soon as it is delivered! Just as its energy performance and operating costs are used to valorize a building today, tomorrow it will be closely linked to the number and quality of the connected services being offered.

The fast development of AI and the strong commitment of GAFAM in this area suggests that the massification of smart aspects in housing, including their DIY (Do It Yourself) solutions, easy to install and concretely not expensive, is now quite close. The ambition to achieve smarter and more enjoyable cities to live in depends in part on the realization of this assertion. Indeed, there is no Smart City possible without Smart Building, in the same way that there is no Smart Building without the Smart Home.

To transform themselves into a Smart City, cities must now invent innovative models to do better with less!

A new value chain and a whole new class of actors providing services are emerging. Whether large groups' subsidiaries or structures originating therefrom, or conversely the *ex nihilo* creation of highly specialized start-ups, everyone will have to find their rightful place without aiming to stack margins by having a global approach. Success will depend at a minimum on breaking away, if not breakthroughs, and in any case, only a "decompartmentalized" approach will allow the emergence of a collaborative and inclusive economic model.

Recent successful crowdfunding examples aim to demonstrate that innovation in project financing is possible and effective. At the city level, one could imagine that the inhabitants themselves become in a way the funders of the smart development of their community... And why not tomorrow with the arrival of virtual currencies?

The topic of Smart Cities and the promises that go along with it will be the outstanding feature of our time! By way of the digital tool, which ought to be used as a real lever of performance of all the issues of territorial policy, we can improve or create cities with more choice, less bottlenecks, increased security, less CO_2, social improvements, and more efficiency in the management of waste, public transport, sustainable development, public lighting, etc.

Data are certainly everywhere but in the form of raw deposits and require to be refined in order to be usable. Open, Big, Smart Data, data management has become a major issue for communities as it is for companies.

Through the generalization of connected objects (sensors, cameras, meters, intelligent lighting, etc.) and the new services that can result from them, neighborhoods, and tomorrow entire cities, will become multi-service-based platforms for the benefit of residents and constituents.

This evolution, or rather this digital transition, represents an opportunity to radically change the way the territory is managed and to optimize costs, to increase the efficiency and quality of services provided and to imagine new ones. We are thus witnessing the emergence of topics such as:

- green energy and saving on bills;
- traffic fluidization;
- the development of new mobility;
- communication with citizens;
- safety assessment;
- improving infrastructure performance;
- improving the quality of life at work and at home.

Feasibility studies and the first POCs (Proof of Concept) have shown that savings made by means of digital technology could cover the investments needed for its development!

Digital technology is also becoming a reality for the Smart City, it must urgently be mastered so that full advantage can be taken of it and in order for the "Smart Territory" to be easy and pleasant to all.

Technically speaking, it has been understood that everything is ready or almost ready, and we are at this pivotal moment when digital transformation is about to transform buildings, neighborhoods and cities by making them truly intelligent and communicative. It is obvious that the new production of buildings and neighborhoods will "embed" these new technologies first and bring out the services that go along with them. The only remaining issue is of the pace of deployment in old infrastructures so that the digital transformation can be massively implemented.

To find out more

Khawam, Z. (2017). L'architecte face au BIM. *Techniques de l'Ingénieur* [Online]. Available at: https://www.techniques-ingenieur.fr/base-documentaire/innovation-th10/industrie-du-futur-outils-numeriques-42669210/l-architecte-face-au-bim-building-information-modeling-c3203/.

Smart Building Alliance (s.d.). Évolution des usages et transition numérique du bâtiment au cœur de la révolution urbaine. Publications de la SBA [Online]. Available at: smartbuildingalliance.org.

Vetil, J.-Y., Flasquin, D. (2017). Le BIM pour la structure – une pratique collaborative. *Techniques de l'Ingénieur* [Online]. Available at: https://www.techniques-ingenieur.fr/base-documentaire/innovation-th10/industrie-du-futur-outils-numeriques-42669210/le-bim-pour-la-structure-une-pratique-collaborative-c3201/.

Éric Pozzo-Deschanel

Architect-urbanist DPLG, Éric Pozzo-Deschanel graduated from the School of Architecture and Urbanism of Versailles and the EM Lyon in management, and specializes in housing design. He is the Director of the Smart Building division of Bouygues Immobilier. He is also Director of the SBA.

Box 5.5. *Smart, did you say smart? Or, how will the Smart Building be the beating heart of the Smart City? (Éric Pozzo-Deschanel)*

Digital technologies at the service of facility safety

Digital transition opens up new horizons in the area of facility safety. The increased quantity of data and their optimized processing have resulted in greatly improving communication between humans and machines. Industrial companies, whether small and medium sized enterprises or large groups, now have access to powerful levers to transform themselves in depth.

The arrival of these new technologies represents a major change for industrialists, different from anything they have experienced so far. Digital technologies enable the improvement of production and human resources including the safety of facilities.

Videos
PSA plant of the future: https://www.youtube.com/watch?v=igW-MyX7PkI

Facility safety, a difficult priority to control

In recent decades, industrial disasters have led to increasing the awareness of all stakeholders in the field. Health, safety and environmental protection are now major issues for both businesses and administration. With this in mind, companies have put in place standards of good practice to meet increasingly stringent regulations and standards. Given the fast growing regulatory and normative complexity, companies are struggling more and more to become compliant.

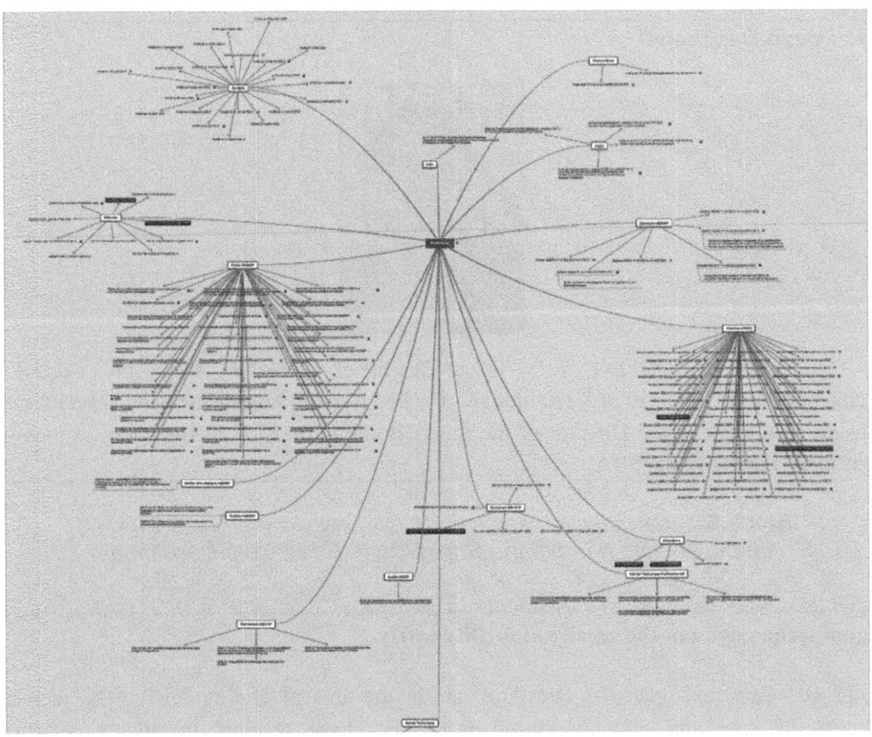

Figure 5.9. *Mind mapping. Number of requirements and interaction for the safety of pressure equipment in France (created by Ealico© on Mindomo©). For a color version of this figure, see www.iste.co.uk/dalpont/process2.zip*

The team responsible for the safety of facilities must maintain up-to-date requirements, normative and regulatory developments, as well as the best practices in the field, regularly collect in-field data and determine what actions they need to implement. This is a long and tedious job, which can be automated and made more reliable by using new technologies.

Reliability and simplicity offered by the cloud

Ealico has developed a new solution to drive compliance of industrial sites. Simplicity and reliability are the cornerstones of using new cloud technologies.

Among other things, the cloud allows information to be centralized and communication facilitated between different systems. With this technology, it is now possible to computerize every requirement in terms of safety and environmental protection (internal, normative and regulatory) and to offer manufacturers a unique experience, both simple and effective, to risk control. The major innovation consists in computerizing business expertise. Therefore, with the Ealico solution, one no longer has to be a regulatory expert to control the safety of industrial equipment. Repetitive and time-consuming tasks have been automated, allowing people in charge to focus on higher value-added tasks.

Figure 5.10. *The Ealico method on PC, tablet and smartphone*

Installing software on the computers of user companies is no longer proposed, instead a single service in the cloud is offered which allows for a standard of flexibility that was previously impossible.

Videos
The Ealico solution in practice: https://www.youtube.com/watch?v=MHeIfTCj33U

Another opportunity offered by the cloud has been to position the company not as a conventional software publisher, but as a digital business specialist. The line between software and service has become more porous. Human coaching has become permanent and with high added value in this digitization process.

As such, digital transformation makes it possible to position humans in the value chain in their rightful place: decision-making based on reliable data is taken wisely. The democratization of cloud solutions has greatly reduced costs for end-customers. The high costs of licensing, maintenance and specific developments have been replaced by a single, much more modest subscription.

Figure 5.11. *Use of the cloud solution on a tablet in a unit (workshop) (source: Neway Partner)*

Artificial intelligence opens up new avenues of possibilities

The democratization of devices connected through the IoT has made it possible to increase the quality and quantity of data stored in cloud solutions, which naturally opens the way towards artificial intelligence.

Developments in the field of artificial intelligence have grown unprecedentedly in the 2010s with the advent of deep learning, which allows machines to improve their own data analytics capabilities to accomplish a defined task. This means that the results of operations calculated by computers are not dictated by humans, but determined directly by the machine.

This technology is already operational. It is particularly used in safety and security to automatically recognize risky situations, objects or people using CCTV cameras and to generate alerts.

In terms of facility safety, the use of data from all industrial sites, whether from real-time sensors, data from equipment manufacturers or operations manufacturers, would provide a much clearer picture of industrial risks. These real-time analysis feedbacks will enable algorithms to predict complex risk-carrying situations in a much more effective manner than anything that has been achieved to date.

Automatic language processing (ALP) technologies are used in the RegTech sector and are particularly capable of recognizing critical requirements among thousands of standards or regulations in record time.

For end-users, everyday life will be easier to manage. They will be assisted by AI which will guide them in their choices and allow them to get answers in real time.

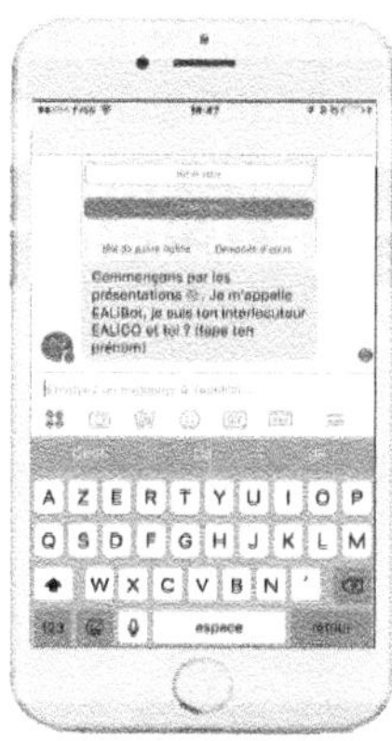

Figure 5.12. *Chatbot. The solution uses AI technologies to automatically provide users with answers concerning industry standards and regulations*

Ealico

Ealico proposes a new way of managing industrial risks. The company has developed a unique technology and services to manage the compliance of industrial sites in a simple and intuitive way.

Labeled Green Tech by the French Ministry of Ecological and Solidarity Transition, the company improves the safety and protection of the environment for its industrial customers in full line with the normative and regulatory requirements of each country. Ealico is therefore a simplified risk management experience.

https://ealico.com/fr

Anthony Martinez

Passionate about the new uses of digital technology and the industry, he is the President and Co-founder of Ealico alongside Valentin Fournier and Armand Gilly.

A graduate engineer from the Saint-Étienne School of Mines, he holds a master's degree in Entrepreneurship and is a graduate of ESSEC and Centrale Supélec.

Box 5.6. *Digital technologies at the service of facility safety (Anthony Martinez)*

How mobile technology can revolutionize health services and public health data in emerging markets

Figure 5.13. *The mobile phone at the service of those who need it*

CommCare is an open source mobile and cloud-based platform designed for non-programmers who want to create and use mobile applications. The decision-making and logical processing platform enables companies to create mobile applications that support high-quality data collection as well as service delivery through customer management, decision support and communication tools for behavioral change.

It is mainly used by community health workers in the most remote parts of the world as a data collection and education tool for patient visits. Supervisors and program managers access the application building platform and the reporting interface through the website that the company Dimagi (see below) operates on a cloud-based server.

CommCare has been specifically designed for remote settings, where no Internet connection is available, and for front-line workers (FLW) who have access to features such as multimedia for low-literate users. CommCare supports J2ME-compatible phones, Android phones and tablets and can capture photos and GPS readings. Besides a myriad of features, CommCare supports multimedia integration (image, audio and video). CommCare mobile versions allow applications to run offline and the data collected can be transmitted to the CommCare website interface when wireless or Internet connectivity becomes available. CommCare also supports SMS, including two-way messaging, broadcast messages and scheduled reminders for each end-user based on configurable data elements such as signup or language preference.

CommCare is the flagship product of Dimagi, a socially responsible U.S. technology company whose mission is to improve the efficiency and quality of services provided to people in need in the most remote parts of the world. Within the context of its work in the healthcare sector, CommCare has helped organizations such as NGOs, ministries, government institutions and businesses in more than 80 countries to develop health information systems and mobile technologies for patient-level disease management, clinical decision support and health system monitoring.

Examples of the impact of CommCare

CommCare has been used in a variety of sectors and has reached the national scale in several countries.

For example, CommCare has been used for mosquito spray campaigns that help protect 52 million people from malaria and other mosquito-borne diseases.

CommCare also supports India's largest digital services program for FLWs worldwide. Led by the Indian Ministry of Women and Child Development, the initiative and its application aim to improve nutritional outcomes at the national scale through "effective monitoring, timely intervention and also as a fact-based decision-making tool".

200,000 FLWs already use the system in India, collecting more than 50 million data per month from millions of beneficiaries.

The social impact of CommCare is measured throughout these projects: the NGO Catholic Relief Services (CRS), which uses CommCare for a Maternal and Child Health Program to reduce infant mortality, has observed that 18 months after the introduction of CommCare, the average number of FLW visits to pregnant women increased by 26%, and 62% more women reported experiencing at least two danger signs at birth.

CommCare has also been used to improve data collection on epidemics such as Zika and Ebola: in West Africa, to combat Ebola, the CommCare-based smartphone-based contact tracking system has enabled real-time identification of contact tracers by time stamping and GPS point collection.

Videos

Introductory video on Dimagi and CommCare:

https://www.youtube.com/watch?v=H-3qxlc7pvA

Overview of CommCare's largest project in India (ICDS-CAS):

https://www.youtube.com/watch?v=pIVu4BCXeCAeature=youtu.be

Shorter introductory video about Dimagi/CommCare in French:

https://www.youtube.com/watch?v=1hOdFb8KQAA

Dimagi

Dimagi is a recognized social enterprise, B Corp and certified Benefit Corporation, with the mission to change the way front-line services are delivered to those in need through better data and better technological tools for the workforce. Its headquarters are located in Cambridge, MA, in the United States, at 585 Massachusetts Avenue, Suite 4, Cambridge MA 02139, USA (info@dimagi.com).

Audrey Philippot-Dahan

Audrey Philippot-Dahan graduated from HEC France and from the Harvard Kennedy School. She joined Dimagi in 2014. She is the Global Director of Dimagi's SaaS business unit, which provides CommCare as a self-service software to hundreds of organizations.

aphilippot@dimagi.com

Box 5.7. *How mobile technology can revolutionize health services and public health data in emerging markets (Audrey Philippot-Dahan)*

LiDAR for sub-marine exploration

From the shallow waters of the Western Australia coral reefs to thermal vents 3900 meters below the surface off the coast of western Mexico, LiDAR has been deployed to improve our knowledge of underwater environments.

LiDAR is short for Light Detection And Ranging. This is similar to RADAR (RAdio Detection And Ranging), except instead of using "radio" waves, light (a laser) is used as the medium to investigate the target. Many of the LiDARs available on the market today are based on Time of Flight (ToF) technology. These ToF sensors use a laser to emit a pulse of light. A simple description is: an accurate timer is used to time how long it takes for the pulse of light to travel to and from a target. This time is then used to calculate the range to the target based upon the speed of light. There is no fundamental limitation on the range of a ToF sensor besides the ability to detect the return photons reflected off the target (ToF LiDARs are used to measure the distance to satellites). Therefore, highly reflective objects can be detected at longer ranges than low reflectivity targets.

There is another class of LiDAR called phase-shift. These scanners emit a continuous beam of laser light that is modulated (either amplitude, frequency, or phase). By calculating the phase of the return waveform in relation to a local copy of itself, the range can be calculated. This technology has a fundamental limit on range (called the range ambiguity), however the newer systems are pushing the range ambiguity out to > 100 meters.

As costs decline, LiDARs are becoming more prevalent and are being mounted on various platforms. Airborne mounted LiDARs are used to map land areas and airborne bathymetric LiDARs are used to map coastal regions down to a water depth of 10 meters. Tripod mounted LiDARs are used for survey and construction work both indoors in factories and outside in plants and construction sites. More recently, automobile and robot mounted LiDARs are used for obstacle avoidance. For all platforms, LiDARs are also used in military applications for target detection, classification, and identification.

One area that has lagged is the use of underwater mounted LiDARs. Due to using light as opposed to acoustics, underwater LiDARs can achieve angular resolutions over 25x better than existing high-end phased array sonars. A search in the literature will reveal that scanning, pulsed, 3D laser imagers have been investigated and deployed underwater over the last few decades for military purposes[16,17,18,19]. Some of these are underwater based and others are

16 McLean, J.W. (1999). "High resolution 3-D underwater imaging", *SPIE Conference on Airborne and In-water Underwater Imaging*, vol. 3716, pp. 10–19.

17 Ulich, B.L., Calmes, L.K., Elkins, W.P. (1990). "Imaging Lidar System", U.S. Patent No 4,964,721.

18 Acharekar, M. (1997). "Underwater Laser Imaging System (ULIS)", *SPIE*, vol. 3079, pp. 750–761.

19 McHale, J. (1997). "Navy moves to Ladar for Minehunting", *Military & Aerospace Electronics*, vol. 8, no. 2, p. 1.

aircraft based to detect underwater mines. In either case, the systems described do not attempt to achieve the range and spatial accuracies provided by current terrestrial laser scanners.

In 2011, 3D at Depth designed and demonstrated the first underwater commercial LiDAR created specifically for the subsea oil and gas industry. The goal of this commercial product was to produce terrestrial laser scanner quality data at an ocean depth of 3,000 meters. Over the past eight years the product has been successfully deployed hundreds of times around the globe to measure and inspect subsea oil and gas assets. This first commercial subsea LiDAR not only attracted the oil and gas sector, but also the scientific community and the underwater archeological heritage sector.

During an October 2018 expedition, the nonprofit Air/Sea Heritage Foundation documented the largely intact wreck of a rare and historically significant US Navy Douglas TBD-1 Devastator aircraft resting under approximately 130 feet of water at the bottom of Jaluit Lagoon in the Republic of the Marshall Islands. Figure 5.14 shows data collection with the subsea LiDAR.

Figure 5.14. *Subsea LiDAR scan of the US Navy Douglas TBD-1 Devastator aircraft resting at the bottom of the Jaluit Lagoon (source: photo by Brett Seymour, courtesy of the Air/Sea Heritage Foundation). For a color version of this figure, see www.iste.co.uk/dalpont/process2.zip*

The goal of the project was to capture precise, repeatable, millimetric data to build an accurate 3D model as a 21st Century site map for this cultural heritage asset; and to serve as a "pre-disturbance survey" for the proposed recovery, conservation and public exhibition of the historic plane at the National Museum of the United States Navy in Washington, DC.

Figure 5.15. *Resulting 3D digital dataset. For a color version of this figure, see www.iste.co.uk/dalpont/process2.zip*

A combination of LiDAR and photogrammetric data was collected. In total, over 92 million LiDAR points and 5.7 million photogrammetric data vertices were delivered. The resulting dataset is shown in Figure 5.15.

Another sector for the technology is the subsea scientific community. In 2013, the Monterey Bay Aquarium Research Institute (MBARI) used a Subsea LiDAR for high-resolution mapping of the seafloor off the central coast of California. The technology was so successful that in 2017 they hired 3D at Depth to develop a custom dual-headed system, capable of operation at a depth of 4,000 meters. In 2018, this system (as well as multi-beam and high-resolution cameras) were used off the coast of Mexico, in the Sea of Cortez, to study deep water hydrothermal vents (3,900 meters). LiDAR data revealed the distribution of deep-sea corals, while the acoustic multibeam sonars only show the seafloor. The higher LiDAR resolution has enabled a new capability: 3D optical mapping of animals that are invisible to acoustic systems.

LiDARs have advantages over other 3D imaging technologies as they can collect more than just 3D data. The same sensor can collect other information remotely such as vibrations, water flow velocities, water temperature and material classification. We expect Subsea LiDAR to follow the same trends as terrestrial LiDAR and be used for applications other than simple 3D data collection. This will be applied not only to subsea oil and gas, but also to offshore wind energy, subsea telecommunications and power cables, port and harbor security, hydroelectric dams and other infrastructure, nuclear power plants, subsea scientific investigations and subsea archaeological studies.

3D at Depth, Inc.

3D at Depth, Inc. is the world's leading expert in commercial subsea LiDAR technology. Advanced subsea LiDAR systems and subsea survey support services help customers transform the value of 3D underwater data. Ranging from data collection and processing to visualization and analysis, 3D at Depth provides accurate, repeatable and millimetric 3D pointed clouds to measure, map and evaluate underwater assets and environments.

Videos

To learn more about underwater LiDAR, below are recent sources of information, data and movies:

http://www.3DatDepth.com

https://annualreport.mbari.org/2018/story/visualizing-the-fine-details-of-the-deep-seafloor

http://www.rovplanet.com/content/magazine (no. 19)

https://mobile.subseaworldnews.com/2019/04/08/3d-at-depth-demos-new-subsea-survey-capacity/

Carl Embry

Carl Embry has been with 3D at Depth since it was founded in 2010 and led the team that delivered the first commercial subsea LiDAR designed specifically for the oil and gas industry. He currently serves as CEO for the company. He received his Bachelors of Science in Electrical Engineering from Memphis State University and his Masters of Science in Electrical Engineering from the Georgia Institute of Technology.

Box 5.8. *LiDAR for sub-marine exploration (Carl Embry)*

The chemical industry operator of process industries in the face of challenges of digital tools: the Adisseo Addixyz system

The operators, working by day or in shifts, are essentially field persons whose main role is to monitor the operation of production equipment, to prevent quality drifts and maintenance needs and to avoid the occurrence of situations dangerous to humans and the environment.

They are key players in the productive system. Often on their own, they have access to extensive information by way of the new digital tools to carry out their job effectively in a situation, while safe and at the same time contributing to their continuous training.

What is the purpose for the manufacturing operator? Becoming self-sufficient on the production lines as efficiently as possible.

Digital tools will enable operators to switch from the notebook, notes and procedures to an interactive tablet with embedded documents (adapted shockproof and explosive atmosphere-proof (ATEX) tablet).

Plant equipment (for example for the sulfuric acid unit: the chemical furnace, electrofilters, thermal recovery unit, the fan, the turbofan, the catalysis case and the conversion, the absorption and chemetics, the washing tower, and the suspended material filter, but this is valid for all units) have been equipped with a QR code that gives access to different documents:

– safety elements (the tablet will indicate hazards and the necessary protective equipment);

– IT diagrams (installation diagrams and equipment connection with nearby equipment);

– virtual tour (relying on shots taken by dual-focus cameras, operators are able to show a tour of the unit as if they were there);

– tutorials, field feedback (if the action requires some experience, the tablet contains tutorials – animated documents, usually in film form – that visually explain the operations to be performed);

– the adjustment mode and the equipment operating mode (in film form but also as tables of values to be obtained for a perfect setting).

A piece of equipment may present a source of danger (temperature, pressure, moving parts, toxicity of the products involved, etc.). The Addixyz tablet will inform the operators of the dangers and risks involved and how to avoid them, in particular by specifying the necessary means of protection as well as the appropriate operations to ensure their safety.

Figure 5.16. *The operator with a tablet examines a piece of equipment*

All the QR codes have been assembled to provide the virtual tour of all the factories from the tablet, allowing the learners the opportunity to learn at their own pace, and in an entertaining and interactive way.

Built with experienced operators, Addixyz is at the heart of practical training as well as companionship. The learner thus discovers the entire production chain through the virtual unit (workshop) and trains for different functions using the 16 mobile learning modules (Addixyz) accompanied by tutored in-field practice.

Before being considered an autonomous operator, he or she is evaluated through a series of quizzes which will automatically be included in the HRIS (human resources information system) in his/her certification file. This certification will be supplemented by a final in-field assessment with the dedicated tutor. This has allowed for a reduction in both the training time and in the rate of equipment failure.

Videos

Operation of the Addixyz tablet in working conditions with the operator:

https://www.youtube.com/watch?v=nGjinZhHdH4

Projet Addixyz Grand Prix Digital Learning Awards Cegos 2018:

https://www.youtube.com/watch?v=nYbgQ6GcwtQ&t=81s

Adisseo

Being a subsidiary of the Chinese group Bluestar, Adisseo, with a turnover of 1.4 billion euros in 2018, is the world's second leader in methionine production (essential amino acid) and holds the number three spot worldwide in animal nutrition with its 2,500 customers. Adisseo's vision, shared by its 2,200 employees, is to provide the planet with healthy, sustainable, affordable and quality food.

Discover Adisseo, leader in animal nutrition and guarantor of the best guarantee for people, the environment and animal production safety:

https://www.adisseo.com/

Georges Scheiber

Georges Scheiber is Director of Transformation and Public Relations at Adisseo. He has held positions of industrial leadership in chemistry, nuclear and pharmaceutical industries.

Box 5.9. *The chemical industry operator of process industries in the face of challenges of digital tools: the Adisseo Addixyz system (Georges Scheiber)*

What does the future hold for engineering?

Industrial engineering includes all of the activities that aim to achieve an investment, that is to translate a process into a production site relying on human and financial resources.

Engineering work typically involves four key steps:

– engineering;

– procurement;

– construction;

– commissioning.

Each of these steps is itself organized into sub-activities. Thereby, the engineering can be broken down into:

– feasibility study;

– preliminary study;

– detailed study;

– execution study.

The accuracy of the project cost estimate increases as the studies progress. Similarly, the planning is increasingly more secure and detailed.

The profession of an industrial engineering company is similar to that of an architect, but it involves more technical disciplines, among which is the manufacturing process and its control (process and process control).

Finally, the challenges of an industrial investment are significant in economic terms (return on investment, date of marketing, control of production costs, etc.) and at the environment level, due to massive (industrial) and sometimes risky production.

This profession has been practiced since the 19th Century at the time driven by the Industrial Revolution, and presents similarities with the ancient construction of large works, some of which date back to antiquity or the Middle Ages (pyramids, Great Wall of China, cathedrals, etc.).

Since its beginning this profession has always been evolving and has had to adapt to the techniques available and to the regulatory frameworks. Yet today, industrial engineering is at the heart of an unprecedented transformation in its history, imposed by the major challenges of our time:

– economic globalization;

– digital transformation;

– the environmental challenge.

In an attempt to shed light on the future of engineering, I propose three paths to be taken, to ask three questions.

What is the industry of the future? What is the expectation of the industrialist of the future? What are the tools of the future of the engineer and engineering?

The industry of the future

What will be the *l'industrie du futur* (France), the *Vier Punkt nul 4.0* industry (Germany), the factory of the future (UK), Advanced Manufacturing (USA), intelligent manufacturing (China)?

This industry will be designed and built quickly because it is imperative to be the first in a market where the lifespan of products is progressively shorter.

Its return on investment will be very short and its CAPEX (capital expenditure) perfectly mastered to meet the growing financial requirements.

It will be intrinsically reliable and/or via contracts that ensure its maintenance in operational condition.

Respectful of the environment, it will be decarbonized, to satisfy the expectations of the population and avoid damaging nature.

It will also be modular and flexible, because it is necessary to be agile to survive an economic world that meets the VUCA (volatile, uncertain, complex and ambiguous) logic. This acronym created by the Army War College in the United States described the conditions resulting from the Cold War. This concept is now applied to the corporate world, subject to the increasingly complex hazards of the regulatory, legal, social, environmental and economic ecosystem. Businesses can no longer draw inspiration from the past. They ought to invent the future.

Finally, it will be small, digital and decentralized. Adidas envisions manufacturing custom-made shoes in its stores using 3D printers rather than mass-producing standard shoes in large, remote factories and shipping them in containers across the seas.

This is indeed a production model that is significantly different from the 19th or 20th Century model.

This industry is already on the move and industrial engineering at the service of the industry of the future is initiating its transformation.

What are the expectations of industrialists of the future? What contract binds them to engineering?

Today, even more than yesterday, industrial clients are financial clients. They wish for a quick answer to two simple questions:

1) What is the price of my production tool?

2) When will I be able to produce?

Nothing original so far, except that the pressure to obtain precise answers to these questions increases, even though the industrial object has not always been accurately described.

The performance of the production tool is considered as being acquired. The main client concern is elsewhere. They are looking for a production and maintenance service capable of producing, the whole accompanied by a contract that limits their industrial risk and allows them to focus on what is essential to them: their product, its marketing and its customers.

We could say then that industrial culture becomes poorer, in favor of a financial culture favorable to the consumption of an industrial object whose costs are to be depreciated over a very short period of time which even makes the idea of rental possible.

Producing a relevant plan is no longer the focus. The industrialist is looking for a partner in engineering in a position to provide a variety of services.

The engineering company needs to broaden its range of skills and integrate:

– financing services;

– production services;

– operational condition maintenance services;

– energy production and distribution services.

The nature of the contracts will therefore evolve significantly towards service contract models similar to what is already encountered in the tertiary sector for a few years. The industrial clients of tomorrow are hoping for services that reduce their industrial risk.

What are the tools of the future for the engineer and in engineering?

> The engineer was one of the main architects of the fantastic progress, which in two centuries, shaped the Earth's surface, changed society profoundly, and transformed our lifestyles, and gave access to space. The engineer as to our current understanding is the culmination of a long history. He makes, embodies the field of scientific knowledge from "learning" to "art" in any facility type, means of production, transportation, communication, and objects produced in series. He cannot do it alone; he has to have a team with the most of varied skills. He needs an organization and resources which he must use properly. (Dal Pont 2012).

Knowledge management

Access to understanding, scientific knowledge, and academic fields has changed in nature. The tremendous leap forward in computing and the research resources associated with them provide means to access it very quickly and accurately. Engineering sometimes holds exclusive knowledge or knows how to search for it quickly. The engineering of the future has access to databases, means for knowledge research and technological monitoring.

Digital models and twins

The management of massive data (Big and Smart Data) and the computing power now available have allowed design tools to make significant progress in only a few decades. This gives engineering the opportunity to expand its sphere of operation.

Therefore, engineering, once design-centered, can now invest in the concretization and maintenance of industrial sites. These three items are driven by means of unique and powerful BIM (Building Information Modeling) tools that enable the management of the entire lifecycle of an industrial asset. This is vertical integration.

Similarly, BIM allows the structuring, exchange of and exploitation of data originating from multiple sources. The different disciplines are coordinated by engineering, and they are brought together within a digital model that is particularly enriched with data issued from the companies in charge of the works and equipment suppliers. The tools make it possible to ensure the compatibility of the facilities to be built and compliance with the design guidelines. For example, the model ensures that piping will be installed without collusion with the metal frame. This is horizontal integration.

Modeling is so close to reality that we are now talking about digital twins.

The first use of these digital twins is to simplify access to information. Every day, plant staff need access to a number of documents (technical records, maintenance history, inspection records, etc.), usually archived in different locations. With a digital twin, this information is integrated into the virtual representation of the production site to which they refer.

Another advantage is that digital twins allow virtual visits to the production site to be performed, thus sparing teams from going on-site to conduct a survey to prepare an intervention, for example.

Virtual doubles also facilitate remote control and/or surveillance operations by linking sensors that ensure the monitoring of production parameters to virtual doubles.

Finally, digital representation can also take into account the operating dynamics of an equipment or production unit. By integrating AI, machine learning and data analysis, the digital twin allows the creation of simulation models that update and adapt to changes.

This learning system is based on:

– data originating from sensors and instruments that describe its operational status;

– human experts such as engineers with in-depth and relevant knowledge of the industrial field in question;

– other similar industrial units;

– broader information but related to the environment in which the system operates.

Digital twins constitute training technology for the industrial Internet of Things, where physical objects can live and interact virtually with other machines and people.

Industrial computing and cybersecurity

This computerized world, from its conception, has a limit that is regularly in the news. Every computer system is vulnerable... The subject of cybersecurity then arises. From transportation and energy networks to Smart Cities including a growing number of plants and production lines, cyber risk is everywhere.

The challenge is therefore both about security and geostrategy. It is part of the governing functions of the state to be concerned about the security of the deployment of a vital network for society and its economy.

While cybersecurity remains everyone's concern, it is up to the ANSSI (French National Agency for Security of Information Systems) in France to help us deal with current and future threats and opportunities.

The future is now and industrial engineering is going through an identity crisis linked to these faster times.

Besides everything that we can state about digital transformation and its effects, the fact remains that humans are responsible for their future. Crisis in Chinese is called *Wei Ji* and combines two ideograms 危机. The first "*Wei*" refers to danger and the second "*Ji*" refers to opportunity.

The engineering of the future is an opportunity to guide a change in the conception of the world. The key point of this change is the consideration of the vulnerability of the planet and its recognition as a living organization.

Éric Colléter

Éric Colléter is CEO of ERAS.

Box 5.10. *What does the future hold for engineering? (Éric Colléter)*

Robots

The industrial company, faced with increased productivity, the flexibility of its manufacture and the increasing diversity of products, needs more and more robots adapted to its trades. Automation has become a necessity.

The term "robot" comes from the Slavic word *robota* which means "work, chore". And indeed, for a long time, robots were intended to replace human beings for difficult tasks (heavy loads, repetitive work, etc.). Today, with the advent of collaborative robots, the goal is no longer to replace human beings, but to assist them. The robot becomes a complementary tool, not a machine taking the operator's job!

An industrial robot is an assembly of mechanical parts and electric motors controlled by a computer. Its design is a subject of mechatronics, a science that combines mechanics, computer science and electronics. It can be equipped with the most diverse sensors, cameras, LiDARs and positioners that make it "smart". Thus equipped, the robot will be able to move, and based on the analysis of images, it will be able to sort objects on a conveyor belt and fetch parts in a store to carry them on a machine tool.

The automotive industry is a major consumer of robots. The list of applications is endless.

Humanoid robots are among the most advanced robots; they can adapt to the person they are helping. Among other things, they use machine learning, which allows them to improve their performance without changing their initial programming.

The Kuka OccuBot (Figure 5.17) is a fully automated measurement system. A six-axis robot controls with maximum accuracy the level of safety and stability of aircraft and car seats under constant mechanical stress. The furniture industry utilizes the OccuBot to control office seats.

Videos
The Kuka application: https://www.youtube.com/watch?v=yWPheSt5lzY

Figure 5.17. *Robot Kuka Occubot*

Figure 5.18 depicts the arm of a Kuka LBRiiwa Robot. One should note, in this arm, the complexity of the integration of the many mechanical and electrical parts, due to the fact that this robot is collaborative and therefore not dangerous even if working close to a human being.

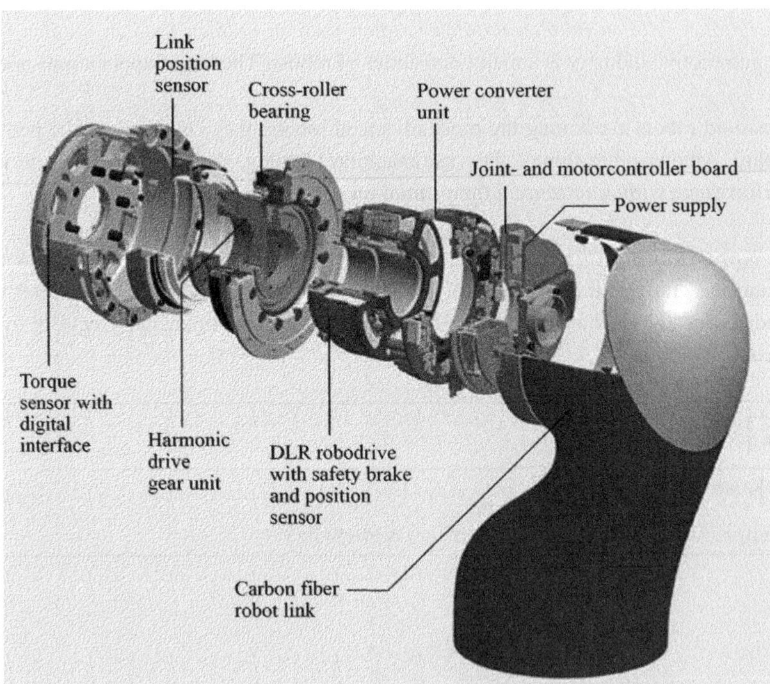

Figure 5.18. *Robot Kuka LBRiiwa. For a color version of this figure, see www.iste.co.uk/dalpont/process2.zip*

Videos

Some examples of application in different areas with Kuka robots:

- robots in the automotive industry:
  https://www.youtube.com/playlist?list=PLcmh-lxe_PW5Wu5kuWuTnog2DphB4 keBR
- robots in the medical field:
  https://www.youtube.com/playlist?list=PLcmh-lxe_PW6eC1Bn7xc9-FozYWlLSFkk
- robots in a foundry:
  https://youtu.be/Zx7W6GgytOs
- sculptor robot:
  https://youtu.be/hgdiXPl3Z0E
- robots in entertainment:
  https://youtu.be/ck8y1sl97BY

Kuka

Kuka is a global automation company with a turnover of approximately 3.2 billion euros and nearly 14,200 employees. As a provider of smart automation solutions, Kuka offers customers everything they need in one place: from components and cells to fully automated systems. Kuka's main markets include the automotive, electronics and general industries, consumer goods, e-commerce/distribution, and healthcare industries. The group is headquartered in Augsburg, Germany.

www.kuka.com

Stefan Dupin

Stefan Dupin is Head of Training at Kuka France. He holds a bachelor's degree in Robotics from the University of Toulon and has been working in the field for more than 20 years. He states: "I am always amazed by the constant evolution of technologies linked to automation and robotics."

Box 5.11. *Robots (Stefan Dupin)*

5.6. References

Abiteboul, S., Peugeot, V. (2017). *Terra Data: qu'allons-nous faire des données numériques ?* Le Pommier, Paris.

Appleby, J. (2010). *The Relentless Revolution: A History of Capitalism*. W.W. Norton & Company, New York.

Attali, J. (2005). *Karl Marx, ou l'esprit du monde*. Fayard, Paris.

Blanc, P. (2018). *L'intelligence artificielle expliquée à mon boss*. Kawa, La Grange Bluffy.

Bouzeghoub, M., Mosseri, R. (eds) (2017). *Les Big Data à découvert*. CNRS, Paris.

CEP Magazine (2018). Artificial Intelligence. *CEP Magazine*, June.

Collins, T., Andler, D., Tallon-Baudry, C. (2018). *La cognition: du neurone à la société*. Gallimard, Paris.

Copeland, J. (2012). *Turing Pioneer of the Information Age*. Oxford University Press, Oxford.

Dal Pont, J.-P. (ed.) (2012). *Processing Engineering and Industrial Management*. ISTE Ltd, London, and Wiley, New York.

Dal Pont, J.-P., Devalan, P., Dombre, E. (2017a). Industrie du futur : outils technologiques. *Techniques de l'Ingénieur* [Online]. Available at: https://www.techniques-ingenieur.fr/base-documentaire/innovation-th10/industrie-du-futur-outils-technologiques-42668210/.

Dal Pont, J.-P., Devalan, P., Dombre, E. (2017b). Industrie du futur : outils numériques. *Techniques de l'Ingénieur* [Online]. Available at: https://www.techniques-ingenieur.fr/base-documentaire/innovation-th10/industrie-du-futur-outils-numeriques-42669210/.

Dal Pont, J.-P., Devalan, P., Dombre, E. (2017c). Industrie du futur : marché, innovation et environnement. *Techniques de l'Ingénieur* [Online]. Available at: https://www.techniques-ingenieur.fr/base-documentaire/innovation-th10/industrie-du-futur-marche-innovation-et-environnement-42670210/.

Espinasse, B. (2017). Introduction au Big Data – opportunités, stockage et analyse des mégadonnées. *Techniques de l'Ingénieur* [Online]. Available at: https://www.techniques-ingenieur.fr/base-documentaire/technologies-de-l-information-th9/gestion-de-contenus-numeriques-42311210/introduction-au-big-data-h6040/.

Julia, L. (2020). *There Is No Such Thing As Artificial Intelligence*. First, Paris.

Magone, A., Mazali, T. (2019). *Voyage dans l'industrie du futur italienne: transformation des organisations et du travail*. Presses des Mines, Paris.

McCartney, S. (1999). *ENIAC: The Triumphs and Tragedies of the World's First Computer*. Walker Publishing Company, London.

Prasad, D. (1996). *Concurrent Engineering Fundamentals*. Prentice Hall, Upper Saddle River.

Rid, T. (2016). *Rise of the Machines: A Cybernetic History*. W.W. Norton & Company, New York.

Ross, A. (2016). *The Industries of the Future*. Simon & Schuster, New York.

Villani, C. (2018). Donner un sens à l'intelligence artificielle: pour une stratégie nationale et européenne. Report [Online]. Available at: https://www.aiforhumanity.fr/pdfs/9782111457089_Rapport_Villani_accessible.pdf.

6

And Tomorrow...

We are living a revolution that has no equivalent in the history of humanity and that, moreover, is a global revolution. It can be rightly argued that the steam engine or the invention of the printing press caused considerable upheaval. This is correct, but these were local events that took decades to spread out.

What is striking about the current revolution is its speed, its globality and also its violence. Two giants are clashing: the United States and China. The United States dominated the 20th Century as a "hyper present country" as well as "hyper-powerful", to quote Philippe Labro (2012).

China, with a market of more than one billion consumers, has gone from copycat to inventor due, in part, to leading universities and a highly centralized power. Two giants, therefore, whose digital societies dominate the world and shape it through their capitalistic and technological power.

The European attitude in this context is worrying. What will happen to old Europe if it does not catch a train that is already running at full speed? What will be the role of Africa, of developing countries in the face of this technological surge? As the saying goes, "no one knows what tomorrow will be made of"; this is currently true!

In the context of this book we try to identify a few key ideas without any pretension. The readers will have to form their opinions for themselves!

Chapter written by Jean-Pierre DAL PONT.

6.1. The beginning of an epic: business, science, technology, the leap forward

In 1975, two American students, Bill Gates and Paul Allen, founded Microsoft, a contraction of microcomputers and software. So many adventures behind this IT company, so many fights, agreements, so many business strategies! **Its Windows software has now become a computer standard.**

The digitization of **analog information** (voice, images, etc.), combined with powerful compaction algorithms, has resulted in huge gains in transmission volumes by telephone lines or airwaves (television).

The discovery of magnetoresistance in 1988 by the Frenchman Albert Fert and the German Peter Gronberg, which won them the Nobel Prize in Physics in 2007, was at the origin of a new field of solid physics: **spintronics**. The latter has multiplied the storage capacity of computer and hard drives by a factor of more than 1,000, thus opening up the possibility to store big data.

Philip Westmoreland examines in his article published in CEP (Westmoreland 2018) the new techniques in chemical engineering that emerged between 2008 and 2018: data analysis, CRISPR, fracking to obtain shale gas, which made the United States independent of fossil fuels other than coal. In 2008, the term cybercrime was unknown and Uber did not exist. The smartphone was in its infancy. The concept of an autonomous car seemed inconceivable.

The author asserts that in 2008, the five largest U.S. companies in terms of capitalization were Exxon Mobil, General Electric, Microsoft, AT&T and P&G. Ten years later, globally speaking, these companies are Apple, Google, Microsoft, Amazon, and Facebook; ExxonMobil ranks 7th. Apple has become the largest company in the world.

We are entering an era of unprecedented uncertainty, where **technology is becoming a weapon** at the service of nations in an **increasingly prevalent economic war**, in which **digital technology** is, by far, predominant.

Digital technology is the main player in globalization.

In the following, we will try to define and modestly imagine the role and impact that AI (artificial intelligence) and digital transition will have at the level of industrial enterprises, their means of production, and on individuals and society in general, at a time when the growth of the world population, metropolization, water stress, impoverishment of or the location of certain strategic raw materials and climate change are all major challenges for humanity. Let us hope it is for the best!

6.2. Artificial intelligence (AI) and economic channels

In his report "*Donner un sens à l'intelligence artificielle*" (Giving meaning to artificial intelligence), Cédric Villani highlights the positive aspects that AI must play in the strategic sectors that he has selected (Villani 2018). The author points out that big data are the very foundation of AI.

AI alters trades, this is inevitable! We shall review a few areas of activity.

6.2.1. *Medicine and health*

AI decrypts mammograms, observes retinas, or scans brains. It has slipped into hospitals and into labs. The medical imaging sector is the perfect cradle for AIs (Vidard *et al.* 2018).

Professor Thierry Philip, president of the Curie Institute, points out, in the Journal of the *Curie Institute* in May 2019, how the analysis of health data by artificial intelligence is extremely promising for research. The **data engineer** at the Curie Institute works on both research data and patient medical records to organize data accumulated over 15 years and make them available to researchers and doctors. It builds a **data warehouse** to gather a mass of information from biological samples, medical reports, chemotherapy data, etc., in one database. Among other things, it is a matter of bringing to light facts that were previously *invisible*, taking into account the problems of storage and anonymization.

The development of a drug costs between one and two billion euros. This is therefore a major risk for pharmacy laboratories. Clinical trials that are essential to the marketing authorization file account for 30% of the total investment. It is hoped that AI-based trials will make reducing costs by about 30% possible, which is significant. It should come as no surprise that Dassault, a world leader in the digital field, which has turned the company into one of the leaders in aeronautical construction, offers its skills to the pharmaceutical industry.

The digital patient is the result of the combination of IT technology, life modeling and medical imaging. It is created by algorithms (Nicholas Ayache in (Vidard *et al.* 2018)). It will help doctors in their diagnosis, in the follow-up and prediction of the evolution of diseases in their clients. It will help the surgeon when operating. It will be a precision medicine, based on sophisticated algorithms.

6.2.2. *The water-energy-food-climate nexus*

This vital issue, in its strict meaning, was discussed in Chapter 2 of the first volume. It is particularly complex, given the multitude of criteria to be taken into account. Powerful computers need to be put to work: meteorology, for example, consumes days of computational resources of the most powerful machines. AI can model regions, entire countries and rivers. Each sector has its own specificities. Box 6.1 reports on water management issues at the time of the transition to digital technology. One should note the importance of new techniques in the preservation, treatment and management of this vital raw material.

Climate change is already having a significant impact on plant design: the increase of a few degrees in cooling water, as is the case with the Persian Gulf, significantly alters the size of heat exchangers. Viticulture needs to rethink its methods: in the future, champagne may be grown in the south of England!

6.2.3. *Intelligent electrical network (Smart Grid)*

Consumer/energy supplier relationships are undergoing changes at a very deep level. At their core can be found the smart meter that allows consumers to manage their needs and the producer to have a real-time viewing of household, professional and city consumptions.

The advent of so-called renewable energy (wind, solar, etc.), which is variable, according to the weather and sunshine for photovoltaics, has complicated the management of the network. The Smart Grid takes generation, transmission, distribution and electricity consumption into account.

The producer at the region or country level is required to bring into service, stop, or modulate fast-starting generators (gas turbines, boilers, etc.) according to the weather. The concept of energy storage is an essential concept where batteries and hydrogen have become major players. This is also true at the plant level (see Box 3.2).

6.2.4. *Artificial Intelligence and Smart City*

The Smart City is also *smart* in its energy management. This is just one aspect of it (see Box 2.1 of the first volume).

Excessive metropolization and urbanization will force giant cities to use AI-based resources to manage their traffic and management flows in the broadest sense, at the risk of seeing, in the strict sense, the asphyxiation of its inhabitants. Who has never dreamed of a *smart, desirable, safe city without traffic jams*?

It is indeed the system city, which we have represented schematically in Chapter 2 of the first volume, that must be taken into account in its entirety with specific expert systems. The exit of 100,000 spectators from a football stadium and who might sometimes be excited can now be modeled to help maintain order; this is just one example among thousands.

Residents will want to know the quality of the air they breathe and of the water that they drink using their smartphones. They want to know where to park. They already do so in a number of cities.

6.3. Artificial intelligence and the consumer

Consumers in developed countries want to know what they have in their glass and on their plate. Increasingly adept at **e-commerce**, they want to know the nature of their food, the origin of the products and their *sustainable development* and *fair trade* components. Therefore, **traceability** has become an essential player in food aspects over the years.

Equipped with their smartphone, consumers can read the label to know the composition and nature of the product that they have in hand, its **authentication**. No product today escapes the concept of **naturality**, which is a term difficult to define: naturality may mean materials of plant origin, bio-sources, etc. But is this a guarantee of a good end product?

Professor Pierre Potier, president of the French House of Chemistry Foundation, used to playfully cite strychnine as a natural product, knowing that it is one of the most virulent natural poisons. Asbestos is also a natural product and so is radioactivity. It comes down to dosage!

Drone delivery is coming. There is no doubt that people with reduced mobility will appreciate it.

6.4. Artificial intelligence, environment and human factor

Who nowadays can do without the 4G smartphone and its multiple features? Who does not take pictures and immediately send them to those around? The world is waiting for 5G, which was the subject of a trade confrontation between the United States and China, where Huawei carved out the lion's share. The ubiquity of these artifacts and their usefulness makes us forget their very heavy **ecological footprint**: a smartphone comprises about 40 metals which are harmful to the environment, originating from mining operations where children often work, not to mention their significant energy consumption involving the respective greenhouse gases.

Today, the global digital system consumes as much energy as a city like San Francisco. Where will this stop, when more and more data centers are being created? Nonetheless, it would be necessary to bring to light what digital systems save in energy, if only in transport. To the best of our knowledge, an evaluation still has yet to be done.

The impact of smartphones, tablets and computers on the user is increasingly being questioned. As new social phenomena, video games and social networks are, for many, becoming an addiction. Their harmful effects on health worry parents and the medical profession: myopia, muscular disorders, disconnection with reality, and psychological disorders are increasingly common. This is even more worrying for children who can no longer get along without these new toys. The human brain no longer rests: it is constantly under pressure due to streams of images, avalanches of emails, and tweets. What space does it leave to reflection?

Who would have predicted the "*gilets jaunes*" ("yellow vests") phenomenon in France, which appeared without any warning in October–November 2018? It is an unstructured movement, not a union movement, therefore it is unregulated, demanding and sometimes causes unheard-of violence. It has mobilized law enforcement and the media for months. The smartphone plays a key role, via social networks, to gather and disseminate information of all kinds, including fake news, and take live photos. Is it a new form of democracy, pushed by public outcry, or an unreasonable demagoguery? How will governments be able to manage this type of situation in the future?

6.5. The human at the heart of the device, at the heart of the system

The theme of the human at the heart of the device, at the heart of the system, comes back as a leitmotif in books and articles from many authors. Obviously, as has already been pointed out, it is the human who creates and not the machine. As such, specifying that humans are at the heart of the system makes sense. However, the question arises as to what they do with modern technologies. Is this for the good or happiness of humanity? This choice is not a new one: Alfred Nobel was able to stabilize nitroglycerin in dynamite, which is an indispensable product for large earthworks, but also a deadly explosive.

6.5.1. *Humans and robots*

What should we think of **robotization** (see Box 5.11)? It is brought forward on the pretext that it eliminates painful, messy, and monotonous trades, which is true.

On the other hand, what will happen to the people that robots will replace and who will become left out of the system? Robotization will enable developed countries to avoid **offshoring**, justified by the low costs of labor in developing countries. Will the fact that these countries are going to run out of work not further widen the gap between rich and poor countries?

The variety of robots is growing, ranging from humanoids to the most modest ones that weed vines or household vacuum cleaners that get the dust where it is hiding!

Humans are social animals, they need **conviviality**. What about the operator surrounded by robots? Telework will avoid transportation problems, but the lack of human warmth usually created around the coffee machine will lead to other problems.

No one knows what will become of **employment**. Clearly, the digital transition is generating skilled jobs, whether they are the new professions of *data officer*, *data scientist*, or expert programmers in increasingly complex algorithms. Repairing a robot will require the intervention of an electrician, a mechanic, and a computer scientist. However, what will be the overall outcome? Will there be work for everyone?

Does working for a certain number of hours, legally set at 35 hours, or at another volume, make sense? Does everyone need to work the same hourly volume both physically, psychologically, or economically? Why would there be adequacy between job offers and availability? Will the digital transition change the game, will it change the very perception of the notion of work? Will humans, so to speak, *freed* from tedious tasks, have the opportunity to choose, to seek other methods of fulfillment?

6.6. System robustness, resilience and fragility

The interconnection of objects and systems leads us into an increasingly complex world, difficult for the majority of humans to grasp. We have already seen the consequences of a power outage in cities such as New York; what will happen tomorrow when an unfortunate bug will put computer systems on their knees? As Box 5.1 makes clear, the human intelligence–artificial intelligence **interface** is difficult to grasp. The question arises of algorithms created by specialists who do not have sufficient knowledge of the field. The two closely spaced crashes of the Boeing 737 Max aircrafts in October 2018 and April 2019, whose analysis is still ongoing at the time of writing, are said to be blamed on the piloting system; this is a complex system that seems to have failed when the pilot attempted to regain control of the aircraft immediately after take-off. Is it true that the pilot could no longer be able to control the machine?

Resilience and **robustness** are key words that the person responsible for a system cannot ignore.

Resilience is originally a term used in physics and metallurgy to define the ability of a body to withstand an impact, to regain its original shape. Let us picture an iron sheet being hit with a hammer: the deformation of the sheet will not be the same when its thickness is one millimeter as it is when it is two centimeters!

The French neuropsychiatrist Boris Cyrulnik introduced this notion into psychology: it can be defined as the ability of an individual to face difficult situations, to overcome them and to move forward. How many children are scarred for life by trauma, including sexual assault? One thinks of burnout, of series of suicides within certain organizations which then begin to decline and become sick bodies.

The resilience of an industrial system can be measured by its ability to withstand a serious accident (fire, explosion), the resignation of a high-level executive, the non-payment of a large bill, the difficult start of a facility while customers wait for the product: it is a long list!

Robustness is a term that speaks for itself: it is the opposite of **fragility**. As we mentioned in Chapter 4, robustness reflects the **confidence** that can be placed in a system knowing that there is no zero risk. To know one's own limits, without engaging in adventures that can prove overwhelming and that can jeopardize the system for which one is responsible, is synonymous with wisdom. Christopher Columbus had prepared his trip well, although he never knew that he had not discovered the route to India!

A system must be **transparent** and **understood by every operator at his or her respective level**. The experts who design it, including computer scientists who are increasingly on the front line, have to become **intelligible**, and forget their densely obscure language.

6.7. GAFA: concerns, fears, myths and phantasms

The acronym GAFA originally referred to the IT companies: Google, Apple, Facebook, Amazon. It became GAFAM when Microsoft was added. In addition to these American companies, one should today consider their Chinese counterparts: Huawei, Alibaba, etc.

They are all characterized by market capitalization and staggering turnovers, billions of customers, gigantic capabilities for data storage, and customer services that multiply as we speak, creating user dependence. Their power appears to be **supranational** in light of their size and their globality: they have no borders. Their taxation is problematic!

Le Monde, on September 6, 2018, announced that Amazon had passed the $1 trillion market capitalization mark, one month after Apple. The capitalization of these two behemoths was equivalent to France's GDP in 2017, which was about $2,290 billion. Amazon's founder Jeff Bezos's fortune was around $167 billion. Still, according to *Le Monde*, his company employs about 550,000 people worldwide, the majority of them low-skilled; packaging and shipping activities are extremely important, especially during Christmas and New Year holidays.

Microsoft has surpassed the $1 trillion mark on the stock market, according to *Les Échos*, on April 26 and 27, 2019, with a share price of $130.59. Microsoft was worth $770 million at the time of the Initial Public Offering in 1986, with a share price of $25.50. Cloud Computing is at the origin of this momentum, under the leadership of its CEO Satya Nadella, an Indian man born to a Marxist father and the father of three children. The *American dream* is still alive.

These companies offer free services, but every click gives them information about who is behind the screen, and they resell this information. Click-paid ***microworkers*** supply GAFA with a variety of information. According to *Le Figaro*, on May 25 and 26, 2019, "click workers" are estimated to be 266,000 in France, 200 million worldwide: is this a new digital proletariat? **The GAFA companies are scaring people!**

6.8. Industrial companies in the face of digital technology

No company can remain indifferent to the digital transition and ignore it. The important question of company bosses is to ensure the survival of the organization that they are in charge of, an organization that may have been entrusted to them by shareholders or of which they are the sole owner. They are faced with a large-scale change management (see Chapter 4) that requires investment in in-house human resources, finding external expertise, and smoothly modifying the system.

It must create new functions at the corporate level, at the plant level, whether they be data scientists or data officers, there is always the fear that newcomers will not be accepted by the old employees and that a lot of time will be devoted to resolving conflicts. As is clearly explained in the study carried out in Piedmont in

Italy (see Chapter 5), it is necessary to have the wisdom to combine the know-how of people in place with the digital agility of young people and experts.

6.8.1. *Cybercrime and uberization*

The first reflex of any company is to safeguard its property and its know-how in the face of cybercrime.

We will leave out *phishing*, ransomware attacks, or attacks on individuals to extort information and money. Harassment on the Internet, which that can assume various forms, and the exchange of illicit information (terrorism, racial hatred, etc.) has no place here.

Cybercrime can be defined as the utilization of the Internet on computers, tablets, and smartphones for appropriating information through illegal manners.

The first intellectual asset that the company wants to protect is the manufacturing process, whereas everything concerning a facility (process diagrams, implementation, etc.) can be stored on a simple USB stick. Raw materials, the nature of catalysts, the list of suppliers and clients are part of the know-how that sometimes took years of work. Computer sabotage, whose sole purpose is to make an information system inoperable, is a real threat at the state level and that of major banking and industrial organizations. Cyberattacks block plants for days or even weeks! Until recently, industrial systems were not designed to withstand these attacks. The manufacturer thought the risk came from within; it can also come from outside.

In France, the **ANSSI** (*Agence nationale de la sécurité des systèmes d'information*, French National Agency for Security of Information Systems) is in charge of helping companies to protect themselves against this new threat.

Uberization can be seen as a kind of cyberattack, although it may be absolutely legal, a point on which we will not comment. Nevertheless, many companies fear this new digital phenomenon.

This neologism derives from the company Uber, which had the idea of putting drivers of passenger cars and customers in direct contact via the Internet. Smartphones, geolocalization and online payment have made this new type of activity possible. Uberization can be recognized as a threat posed by customer piracy.

The uberization of the economy in the hotel industry, transport, technical support, to name but a few sectors of activity, offers multiple advantages for customers but raises multiple controversies, on the one hand, among the traditional companies that find themselves dispossessed of their clients and, on the other hand, among protagonists without social protection, reduced to certain precariousness.

France is entering the era of *cyberwarfare*, declared Thierry Breton, the boss of IT company ATOS, to the French newspaper *Les Échos* in July 2019.

6.8.2. *Software hybridization*

Michael Valentin (Valentin 2018) quotes Marc Andreessen Horowitz: "In short, software is eating the world." Obviously no one can escape hyperconnection and applications of any kind. The author calls this invasion of digital technology in all sectors of the economy, including the productive system, "software hybridization". The question is to "hyperconnect" humans, machines and products.

Chemical industries were the first to embrace digital technology, to instrument, and to equip themselves with control systems. Manufacturing industries, which are lagging far behind, are wholeheartedly adopting automation and robotization, as part of a strategy that many see as an essential prerequisite for survival.

The "hybrid cloud" system consists, for a company, of using the cloud as well as in-house means and outsourcing others.

This profusion of software programs and scalable digital systems leads the company to problems of architecture, connection, and interoperability.

This is reminiscent of the introduction of quality systems some 30 years ago: ignoring the principle of **subsidiarity** has cost some companies dearly. Subsidiarity consists of informing and having tasks performed at the closest level to the persons concerned, without considering this as being derogatory at the *lowest* level of the organization. A plant manager is not supposed to know about maintenance work orders every day.

6.8.3. *After Fordism and Toyotism, Teslism?*

We described Fordism and Toyotism in the first chapters of the previous volume. These two industrial systems, both from the automotive industry, have been the archetype of industrial organizations 50 years apart.

By an ironic twist with a familiar fate, Toyotism, which empowered workers, in contrast to Ford's Taylorism in 1907, that enslaved them to the assembly line, emerged victorious from the U.S.-Japan confrontation. More than 50% of American cars are now of Japanese origin, while Japan has very few natural resources. The United States discovered Toyotism in 1990 (Valentin 2018, p. 44).

That shows the importance of the concept of *manufacturing* in the success of the company. In another twist of fate, Toyotism is based mainly on a few industrial management tools and not on IT technologies, although the robot is the supreme master of Toyota lines.

Elon Musk, iconic entrepreneur, boss of Tesla and SpaceX, is at the heart of Michael Valentin's book, rightly focused on Tesla's successes. This did not prevent him from exposing the setbacks.

The author describes the seven principles of Teslism (Valentin 2018, p. 36). Among these, "cross-integration" is paramount. Very succinctly, it is based on the following four levels:

– **strategic integration**: vertical integration, including that of clients and suppliers;

– **organizational integration**: horizontal, it is necessary to decompartmentalize;

– **technological integration**: cross-cutting with hybridization of traditional and digital businesses, including change management. This is not dissimilar to the Italian experience (see Chapter 5 of this volume);

– **peripheral integration**: social and environmental.

The author concludes that "Teslism is not an end in itself, but a a major source of inspiration to enter the fourth age of industry". The responsibility for these remarks is obviously left to the author. Moreover, what novelty Teslism brings is legitimately questionable and one should try to distinguish between what is real and what is media. Tesla's future may be the answer to this question!

6.8.4. *Business and governance: products*

The company can no longer hide much. This is perhaps one of the positive aspects of globalization and social networks: there are no secrets. The industrial company ought to be virtuous because its products can be "seen". This is the case of chemical industries, for which the effects of products are sometimes indiscriminately evaluated by a hyperconnected consumer, turned digital actor because of his or her smartphone!

The case of glyphosate is a reference: marketed in 1974, this herbicide is non-selective, in other words, it destroys *everything* in its path. It is the world's top seller and looks irreplaceable, at least economically. It has been established that "organic" crops have yields well below those of "non-organic" crops; what should be read is that they make use of pesticides. Mechanical weeding is expensive, if not impossible, in rough terrain. Farmers who are banned from using glyphosate, the "miracle" product, complain loudly and clearly about unfair competition from their colleagues who are allowed to use it. At the root of this problem, at least in France, is whether the product is really harmful. Is it carcinogenic? If so, at what dose? Laboratory trials are multiplying, but the public does not trust them. There is confusion between danger and risk.

The traceability of products, the assessment of their impact on the environment in the broader sense, and their governance (*product stewardship*) have become essential; the company cannot avoid it and the tools of the digital transition are there to help. Governance, the subject of ISO 26000, requires the company to adopt a sustainable attitude; it has become the law in some countries, such as France.

Danone, the global agri-food giant, perhaps represents the archetype of a change that is taking place among multinationals. Its new boss Emmanuel Faber says that "biodiversity is part of global food security. Every consumer must be able to become sovereign of his or her food again". Capitalism must "create value for shareholders and for society" (*Le Figaro Magazine*, April 13, 2018).

Danone's U.S. subsidiary has been awarded the new B Corp label. This certification, based on 200 criteria, created in the United States (a capitalist country if any) in 2006, is sought by companies that want to join a citizen-based capitalism.

6.8.5. *The chemical engineer, the project management*

Chemical engineers play a key role in the industrialization of the process where they constitute a pivotal function between research and development and project engineering. We mentioned this in Chapter 3. **They are at the heart of the system!** They are architects, *orchestral conductors*. They define the process, and establish heat and material balances. They provide a wealth of information to the project team that can include about twenty different trades in chemical industries.

Nevertheless, digital technology emerges with its new tools, explained in Chapter 5. The tool box, as the essential baggage of technicians and the guarantor of their added value, is expected to understand what will turn a unit or a plant into a so-called tool of the future, a computerized tool. Engineers of yesteryear specified the instrumentation; they must now participate in the definition of the algorithms at the base of the hybridization of the system. Will they have to define robots? If so, which ones and up to what level? They will have to work with mechatronics engineers. They are responsible for contributing to defining the digital twin and for listing the operating variables to follow, the alert thresholds, etc.

What is certain is that project management is radically changing. Instead of a certain *linearity*, there is a shift towards pooling the project among all its actors, namely, its stakeholders. As described in Box 5.10, the project can be viewed from the beginning by means of a virtual model. Each supplier can intervene at its own level. Coordination, project cohesion, the diffusion of responsibility, and the weakened authority of the project manager are now questionable. And yet, if things do not go according to plan, the project manager will be the sole person responsible.

6.9. Towards a Black Box Society?

Our society will depend more and more on information technology with smartphones: everything is connected and everything will be increasingly connected. Everything can be known, it is a matter of algorithms. Frank Pasquale (Pasquale 2015) questions the **opacity**: every click reveals information about us and no one knows where it will be stored; all this will be used to influence our vote as citizens or our consumer choices one day.

There is no algorithmic ethics nowadays! Who can guarantee that the household robot is not spying on the whole household? *Smart* electric meters indicate the presence or absence of inhabitants. Do we really know the final destination of this information? Some countries are already giving scores to their citizens. In the country that invented the *Habeas Corpus*, English people are photographed 10 times a day on average as soon as they leave their home.

Is digital transition a threat to democracies? Shall we be able to counter the power of GAFA?

We are experiencing a digital coup d'état, exclaims Laurent Alexandre. Will states be able to withstand GAFA? The title of his book "*L'IA va-t-elle aussi tuer la démocratie ?*" (Will AI also kill democracy?) is evocative.

6.10. Conclusion

Can we make an assessment of a system that is in complete evolution, in motion, in acceleration? This is a difficult task, but is nevertheless essential. We shall take the risk!

Surprisingly, the digital revolution has brought *manufacturing* to the forefront. Even in the 1980s, authors such as Hayes and Wheelwright (Hayes and Wheelwright 1984; Hayes *et al.* 1988) drew America's attention to the fact that it was becoming a "Casino Society", that a business was merely a "capitalistic merchandise" that was bought and sold according to the stock exchange. Manufacturing was considered a "dirty job"; it was not raising any interest among the best students, who were instead attracted by consulting or law. We have witnessed where this has led the American auto industry, which has been deeply undermined by its Japanese competitor!

The success of the industrial enterprise begins on the shop floor: Toyota understood it after the war. Manufacturing is synonymous with employment. Detroit, once a thriving city, is looking for a second wind.

Chemical industries offer a huge variety of industrial tools; it all depends on the industry under consideration, on continuous or batch-based operations, the size of the tool, its technology, its location, or the markets, etc. Plants that implement solid materials have very different characteristics from those that implement fluids, which are easier to domesticate. Oddly enough, the manufacturer, which is discontinuous in essence, is very similar to batch industry.

What all industries share now is an imperative need to implement digital hybridization, both from a managerial and operational point of view. **The digital revolution cannot be ignored**. Special attention is given to its implementation.

Another necessity is societal acceptability, environmental respect, knowing that the **job-growth-ecology-economy nexus** is the source of multiple contradictions. We want to stop nuclear power plants in operation, in order to move to renewable energy, including wind and solar power, without specifying the number of wind turbines and the surface area of equivalent voltaic panels, or the amount of resources to be mobilized, and without making a serious CO_2 assessment. The enormous cost and the duration of dismantling a nuclear power plant is simply overlooked. Some countries are stopping their nuclear power plants and restarting their coal-fired power plants! Hydrogen is highlighted, forgetting that its production by SMR produces CO_2 (see Chapter 2). Pollution is relocated! The issue of batteries for electric cars is not always addressed in a professional way.

The concepts of the sharing economy (where the concept of a fab lab is perhaps a beginning of a circular economy), Smart Buildings and Smart Cities will become unavoidable, driven by demographics and climate change that seems to be very real. The low water levels of the Rhine are severely affecting the powerful German industries that were created on its shores, weaned off raw materials and that increasingly have difficulties selling their products. Will the phenomenon continue, improve, or worsen?

New forms of industry are to be invented: is this the end of gigantism? Are we going to relocate the means of production close to consumers? Are we going to build factories, paradoxically, in the middle of the city? This is already the case, without the man on the street noticing it. The race for decarbonization is on!

Never before has the need for **systemic analyses**, conducted by panels of technicians and citizens with appropriate **metrics**, been greater. To this end, Big Data, the new black gold, according to some, and algorithms will help (Abiteboul and Peugeot 2017). Simulation and optimization techniques have a bright future ahead of them.

The fact that we must "think of humans differently" based on Big Data is the subject of Gilles Babinet's book (Babinet 2015).

While the digital revolution was making its first appearances, Peter Stenge was advocating a "necessary revolution"[1] and showed "how individuals and organizations are working together to create a sustainable world"[2] (Stenge 2008). Has the world become more sustainable in ten years? What is certain is that an awareness has taken shape and that the COP 21 has contributed to it.

Are citizens allowed to speak openly, at the risk, if not, of seeing "yellow vest" movements spread, not without unjustified excesses? Our viewpoint applies to developed and countries, to rich countries. What is the viewpoint of developing countries, or poor countries?

It has been said, for decades, that the time has come for engineers; it seems that this is now true. If the technical bases that were taught to them not so long ago remain the same (and they should have assimilated them well), the digital transition forces them to adapt to a new state of mind. They always knew how to do it, they will do it.

1 "Necessary revolution."

2 "How individuals and organizations work together to create a sustainable world."

6.11. Box

Water management

The major challenges of the digital transition

The world's population is expected to reach 9 billion in 2050, with increasing urbanization (34% of the population in cities in 1950, compared to 54% in 2015, and 66% expected in 2050). For example, needs should increase by 50% for food, 45% for energy and 30% for water by 2030, compared to 2012.

With unevenly distributed water resources, the challenges remain enormous in allowing the entire population access to clean water and sanitation as well as limiting pollution. All over the world, territories are facing the impacts of climate change, suffering extreme events with sometimes dramatic consequences (floods, droughts, etc.).

Major technological breakthroughs have occurred over the past 10 years, brought by computing for the capacity of processing information, by microelectronics for the development of ever more efficient and less expensive sensors, or by telecommunications through advances in information transfer and, especially, wireless.

Recent advances in the digital field are opening up broad perspectives in water management methods, whether for developed or developing countries, and allow solutions to be considered to respond to developments in the years to come.

Therefore, the intelligent management of water will enable, using new information and communication technologies, controlling and diagnosing problems, controlling processes, prioritizing and managing, continuously and remotely, maintenance operations and utilizing the data provided to optimize all aspects of the performance of water management.

Production and distribution of drinking water

8% of the water collected at the resource level is employed for the production of drinking water.

The UN estimates that 1.5 to 2 billion people worldwide lack safe drinking water at certain times of the year. Population and economic growth, combined with declining reserves due to climate change and resource pollution, will further exacerbate the difficulties in supply in the coming years.

Conventional water sources, such as precipitation, melting snow and runoffs captured in lakes, rivers and aquifers, are no longer sufficient to meet human demand. The use of unconventional water sources – desalination, reutilization of wastewater – becomes essential

(Jones *et al.* 2019). Digital developments can also be tools to limit consumption and preserve the resource.

Unconventional water sources

Desalination (Rovel 2017)

21 billion m^3 of water are produced each year by desalination (0.45% of total consumption). Desalination allows fresh water to be obtained from brackish or salty water. It is growing rapidly: the world's installed capacity has increased by 160% over the past ten years. The two most commonly used processes are distillation (40% of desalination capacity) and reverse osmosis, which present better energy performance and whose growth rate is the highest.

The release of brines into the natural environment, for their part, are estimated at 142 million m^3 per day and improved management strategies are needed to limit negative environmental impacts and reduce the economic cost of disposal.

Figure 6.1. *Desalination station in Djerba (Tunisia)*

Wastewater reuse

Faced with the risk of water scarcity, wastewater reuse constitutes a key area, under study by many countries for the irrigation of green spaces or crops, firefighting, road washing or for recharging groundwaters.[3]

The example of Singapore is the best known. Singapore imports 30% of its water from Malaysia (an agreement agreed to last until 2060). The city-state is striving to reduce its dependence by developing new technologies for recycling (NEWater), desalination, or rainwater recovery. Waste water undergoes state-of-the-art treatment combining microfiltration, reverse osmosis and UV radiation, before being mainly reused in industry, but also partly for direct consumption. Acceptance by the general public has required heavy communication and information campaigns.

In addition to the barriers associated with acceptance, use for direct consumption may be limited by high salinity and the presence of micropollutants in these waters. Research is underway to combine physical and chemical treatments and obtain the qualities required for these waters (Gazette 2019).

Distribution (Hydreos 2016)

In the field of drinking water distribution, major developments have emerged in recent years with the use of digital technologies in:

– smart metering (remote monitoring and remote meter reading): the generation of consumption data at a high spatial resolution (every household) and at significant time frequencies (from a few hours to a few minutes) opens up perspectives for the characterization and modeling of networks, but also allows for the development of strategies for managing water use with end consumers (Cominola 2015). It can thus act on two levers to preserve the water resource: reduce consumption and identify water losses;

– real-time monitoring of water quality. Its implementation serves three main objectives: to ensure consumer protection from malicious acts, to improve supplied water quality and to optimize the operation of water treatment facilities and distribution systems;

– leak detection: worldwide, one-third of communities report a loss of more than 40% in drinking water due to leaks in the system (Mounce *et al.* 2015). Communities are increasingly using active control to limit losses, with a a policy of systematic research (acoustic sensors, sectorization, pressure monitoring), leak localization and immediate repair;

– the knowledge of heritage and its long-term evolution: the advance in wireless detection technologies and the reduction in their cost now allow for dense network coverage and thus real-time infrastructure monitoring. By combining these data with appropriate processing

3 www.brgm.eu.

techniques, it is possible to improve the prediction of defects and the management of installations;

– real-time plant and network management: the models used for predicting facility operation have long relied on "hypotheses" data, collected from previous events, providing a unique response to a study case. With the development of online measurements, it is possible to generate instant simulations, taking into account the evolutionary parameters. Therefore, the analysis of the data currently carried out has gone from descriptive analysis (transformation of data into knowledge) to predictive analysis (modeling of future events) due to machine learning techniques and moves towards prescriptive analysis: the study of the impact of different measures on performance and orientation towards the optimal solution.

Wastewater treatment

Communities are facing strong challenges that drive them to evolve in the management of their sanitation systems:

– water resource preservation: optimized management of wastewater collection is essential to limit discharges into the natural environment and thus impacts on the quality of water bodies (impacts on the aquatic ecosystem, impacts on groundwater quality, etc.);

– flood control: a large part of the collection network is still in unit operation and overflows during heavy rains can lead to floods having human, material and environmental consequences due to the discharge of pollutants into the natural environment;

– optimization of energy consumption: wastewater treatment plants are generally very energy-intensive and account for a significant share of energy expenditure related to sanitation, which increases the environmental footprint of this activity;

– infrastructure management: detection and reduction of leaks, detection and reduction of parasitic clear water, network maintenance, reduction of olfactory nuisances, etc.

In the face of these challenges, innovations are multiplying:

– treatment plant of the future (see video below), with a trend towards recovery plants for effluent resources: recovery of energy present in effluents, fertilizer production, treated water reuse;

– the use of fiber optics for network monitoring in order to diagnose, alert, and quickly locate network problems and thus avoid potential incidents;

– the identification of pollution events ahead of the treatment plant and the implementation of a proactive, real-time response;

– real-time flow management, coupled with weather forecasts, to control the arrival of effluents towards the treatment plant or buffer tanks;

– the simulation of effluent treatment to reduce untreated releases, energy costs and investments.

Videos

The purification plant of the future:

https://www.youtube.com/watch?v=8wRihD0nBtk

Water in chemical industry

In terms of water resources, with a consumption of 800 billion m^3 in 2009 and an estimate of 1,500 billion m^3 in 2030, the industry is the second largest consumer of water in the world after agriculture. The world's water resources are decreasing and prompting concrete action to limit consumption. The plant is evolving in this direction to identify production methods that are less water- and energy-intensive.

Water is present throughout the processes, it can be used to wash, rinse, dissolve, dilute, heat, or cool. It can, under certain conditions, be used as a raw material for the manufacture of the product, sometimes with very specific specifications (drug industry, agri-food, etc.).

Currently, effluents in the plant are commonly mixed to be treated at the end of the chain, without knowledge of the origin of consumption and pollution.

In reflection about the plant 4.0, the treatment of effluents must find its place, with the implementation of mapping and diagnostics to understand the water pathway within the plant and optimize the treatment at different points as well as recycling or reuse at the heart of the process.

Water-energy-food nexus[4]

Figure 6.2a. *Water–agriculture relationship: rice paddies in Nepal*

4 http://www.fao.org.

Figure 6.2b. *Water–agriculture link: rice paddies in Nepal (continuation)*

These three sectors are closely linked:

– agriculture accounts for 70% of the world's total freshwater withdrawals;

– 90% of the energy currently produced is water-consuming;

– agriculture and the food chain account for 33% of the worldwide energy demand.

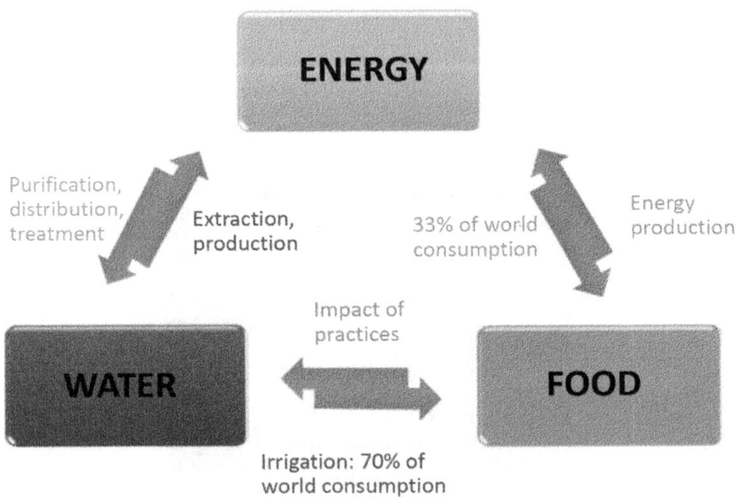

Figure 6.3. *Example of water-energy-food interactions. For a color version of this figure, see www.iste.co.uk/dalpont/process2.zip*

The inextricable links between these topics require integrated approaches to ensure food and water security, sustainable agriculture and energy production. The water-energy-food safety nexus approach can help achieve integrated resource management and policy planning (International Renewable Energy Agency 2015).

Surveillance of the natural environment

In Europe, the EC 2000/60 framework directive for water requires the achievement of target results on the state of water bodies. The procedure comprises an assessment of the chemical and ecological status, including a number of parameters to be measured and thresholds not to be exceeded. Measurements are usually taken by laboratory samples and analyses, but sensors are also available and innovative monitoring tools are coming to the market:

– surveillance drones to collect samples and carry out analyses (example of the drone developed by IFREMER: see video below);

– passive samplers: positioned in the medium to be analyzed, these samplers concentrate pollutants and even allow the analysis of substances present at the trace state;

– use of satellite data to define the quality level of water and dams in hard-to-reach areas (Cochonneau *et al.* 2013), but also for flood monitoring and prevention[5];

– biosensors: comprised of a biologically active element (living cells, fish, etc.), they are sensitive to environmental pollution and their reaction is transcribed as an electrical signal.

Videos
Drone developed by IFREMER: https://www.dailymotion.com/video/x3faoy0

Challenges of the introduction of digital technology

Data management

The evolution towards Smart Water results in an increase in the amount of data. The collected information represents a significant wealth but may also emerge as an obstacle to the adoption of smart technologies. For example, some communities are reluctant about the topic for fear of being overwhelmed by the amount of data. The challenges lie in the volumes collected, the variety of formats and the speed of collection.

5 http://www.esa.int/fre/ESA_in_your_country/France/La_cartographie_des_inondations_par_satellite_au_service_de_la_securite_civile_dans_l_Est_de_la_France.

It is necessary that data be widely shared between potential users (users, research laboratories, government, suppliers, etc.) and between services (water, energy, etc.) to develop innovations in the future.

Cybersecurity

In recent years, hacker attacks have endangered various environmental infrastructures. To protect yourself, it is necessary to establish an analysis of one's own situation and to identify one's weaknesses. The risk must be estimated realistically, avoiding disaster scenarios.

All stages of the value chain (processing, distribution, commercial service) must be secured.

Behavioral evolution

The deployment of technologies still faces obstacles related to the very rapid evolution of the sector, out of step with how communities function, which are rather conservative, particularly due to the aging of the workforce.

Concerns are raised by the capabilities of managing large amounts of data, which require changing tools and methods, as well as by the risk of rapid obsolescence of digital solutions. Problems with isolated systems and the lack of interoperability are also holding back investments in the sector.

Trends for the next years (Cominola* et al. *2015).

The world of water is becoming digital. Sensor and data-related technologies are largely installed in networks and the trend is expected to intensify in the coming years.

In the near future, three technologies are expected to make a significant impact:

– gamification, or the utilization of mechanisms from games, to test the results of intelligent solutions and raise public awareness to resource preservation;

– improved battery performance in communication systems;

– data analysis with artificial intelligence and machine learning tools.

A refocusing of water services towards their core business, water and sanitation management, is expected to occur in the coming years. Professions will evolve with, in particular, the necessary use of data scientists in teams and/or partnerships, such as subcontracting ones, which should develop around data-related subjects and communication technologies with solution providers.

For further information

Cochonneau, G., Martinez, J.-M., Cerqueira Condé, R., Calmant, S., Seyler, F., de Souza, E., Le Hersan, E. (2013). Chaînes d'analyse et de traitement de données satellitaires pour produire des séries temporelles de niveau et de qualité des eaux de fleuves et barrages [Online]. Available at: http://devlog.cnrs.fr/_media/jdev2013/ore-hybam_cochoneau_poster_jdev2013.pdf?id=jdev2013%3Aposters&cache=cache.

Cominola, A., Giuliani, M., Piga, D., Castelletti, A., Rizzoli, A. (2015). Benefits and challenges of using smart meters for advancing residential water demand modeling and management: A review. *Environmental Modelling & Software*, 72, 198-214.

Hydreos (2016). Gestion intelligente des réseaux d'eau – Quels besoins ? Quelle réalité ? Quelles perspectives ? Smart Water Study.

International Renewable Energy Agency (2015). Renewable energy in the water, energy and food nexus [Online]. Report. Available at: https://www.irena.org/documentdownloads/publications/irena_water_energy_food_nexus_2015.pdf.

Jones, E., Qadir, M., van Vliet, M.T.H., Smakhtin, V., Kang, S.-M. (2019). The state of desalination and brine production: A global outlook. *Science of The Total Environment*, 657, 1343–1356.

Mendret, J. (2019). Réutilisation des eaux usées : Quels sont les pays les plus en pointe ? [Online]. *La Gazette des communes*. Available at: https://www.lagazettedescommunes.com/612777/reutilisation-des-eaux-usees-quels-sont-les-pays-les-plus-en-pointe/.

Mounce, S.R., Pedraza, C., Jackson, T., Linford, P., Boxall, J.B. (2015). Cloud based machine learning approaches for leakage assessment and management in smart water networks. *Procedia Engineering*, 119, 43–52.

Rovel, J.M. (2017). Dessalement de l'eau de mer [Online]. *Techniques de l'Ingénieur*. Available at: https://www.techniques-ingenieur.fr/base-documentaire/environnement-securite-th5/procedes-de-traitement-des-eaux-potables-industrielles-et-urbaines-42318210/dessalement-de-l-eau-de-mer-w5700/.

Hydreos

Hydreos, the Grand Est region water cluster, is a structure serving members, which are players in the water market. By creating synergies to promote the development of innovative projects, *Hydreos* supports its members in implementing actions to promote the sustainable improvement of water in all its forms.

In 2019, the three French water clusters, *Hydreos*, *Aqua-Valley* and *Dream Eau et Milieux*, have come together and founded the national competitiveness cluster of the water sector, PCF Eau, which covers the entire water sector throughout the whole of France.

Sophie Altmeyer

As a chemical engineer and graduate of the *École nationale supérieure de chimie* in Rennes, Sophie Altmeyer was, from 2000 to 2011, a research engineer and then director of the PROGEPI, the laboratory transfer center of the *École nationale supérieure des industries chimiques*. Since 2011, she has been project manager and then technical manager at *Hydreos*.

Box 6.1. *Water management (Sophie Altmeyer)*

6.12. References

Abiteboul, S., Peugeot, V. (2017). *Terra Data. Qu'allons-nous faire des données ?* Le Pommier, Paris.

Babinet, G. (2015). *Big Data. Penser l'homme et le monde autrement*. Le Passeur, Paris.

Hayes, R., Wheelwright, S. (1984). *Restoring our Competitive Edge. Competing Through Manufacturing*. John Wiley & Sons, New York.

Hayes, R., Wheelwright, S., Clark, K.B. (1988). *Dynamic Manufacturing. Creating the Learning Organization*. The Free Press, New York.

Labro, P. (2012). *Mon Amérique. 50 portraits de légende*. La Martinière, Paris.

Pasquale, F. (2015). *Black Box Society. Les algorithmes secrets qui contrôlent l'économie et l'information*. FYP Éditions, Limoges.

Senge, P. *et al.* (2008). *The Necessary Revolution. How Individuals and Organizations Are Working Together to Create a Sustainable World*. Doubleday, New York.

Valentin, M. (2018). *Le modèle Tesla. Du toyotisme au teslisme : La disruption d'Elon Musk*. Dunod, Paris.

Vidard, M., Bloch, L., Joffrin, L. (2018). L'IA au cœur de l'humain. *Libération*, special issue, 2.

Villani, C. (2018). Donner un sens à l'intelligence artificielle. Pour une stratégie nationale et européenne [Online]. Report. Available at: https://www.aiforhumanity.fr/pdfs/9782111457089_Rapport_Villani_accessible.pdf.

Westmoreland, P. (2018). Revisiting the Future of Chemical Engineering. *CEP Magazine*, October, 26.

List of Authors

Philippe ARPENTINIER
Air Liquide
Jouy-en-Josas
France

Laurent BASEILHAC
Arkema
Paris
France

Marie BASIN
Air Liquide
Jouy-en-Josas
France

Jean-Pierre DAL PONT
SECF
Paris
France

Marie DEBACQ
Cnam – AgroParisTech
Paris
France

Raphaël FAURE
Air Liquide
Jouy-en-Josas
France

Matthieu FLIN
Air Liquide
Jouy-en-Josas
France

Philippe JACQUES
University of Liège
Gembloux
Belgium

Vincent LAFLÈCHE
École des mines de Paris
France

Diana TUDORACHE
Air Liquide
Jouy-en-Josas
France

June C. WISPELWEY
AIChE
New York
USA

Index

C

D

E

M

N

O, P

T

U

V

W, Y

Summary of Volume 1

Chapter 4. Chemical Engineering: Introduction and Fundamentals

Marie DEBACQ, Alain GAUNAND and Céline HOURIEZ

Chapter 5. Chemical Engineering: Unit Operations

Marie DEBACQ

Other titles from

ISTE

in

Chemical Engineering

2020

BELAADI Salah
Thermodynamic Processes 1: Systems without Physical State Change
Thermodynamic Processes 2: State and Energy Change Systems

SCHAER Éric, ANDRÉ Jean-Claude
Process Engineering Renewal 1: Background and Training
Process Engineering Renewal 2: Research

2019

HARMAND Jérôme, LOBRY Claude, RAPAPORT Alain, SARI Tewfik
Chemostat and Bioprocesses Set
Volume 3 – Optimal Control in Bioprocesses: Pontryagin's Maximum Principle in Practice

2018

LOBRY Claude
Chemostat and Bioprocesses Set
Volume 2 – The Consumer–Resource Relationship: Mathematical Modeling

2017

HARMAND Jérôme, LOBRY Claude, RAPAPORT Alain, SARI Tewfik
Chemostat and Bioprocesses Set
Volume 1 – The Chemostat: Mathematical Theory of Microorganims Cultures

2016

SOUSTELLE Michel
Chemical Thermodynamics Set
Volume 5 – Phase Transformations
Volume 6 – Ionic and Electrochemical Equilibria
Volume 7 – Thermodynamics of Surfaces and Capillary Systems

2015

SOUSTELLE Michel
Chemical Thermodynamics Set
Volume 1 – Phase Modeling Tools
Volume 2 – Modeling of Liquid Phases
Volume 3 – Thermodynamic Modeling of Solid Phases
Volume 4 – Chemical Equilibria

2014

DAL PONT Jean-Pierre, AZZARO-PANTEL Catherine
New Approaches to the Process Industries: The Manufacturing Plant of the Future

HAMAIDE Thierry, DETERRE Rémi, FELLER Jean-François
Environmental Impact of Polymers

2012

DAL PONT Jean-Pierre
Process Engineering and Industrial Management

2011

SOUSTELLE Michel
An Introduction to Chemical Kinetics

2010

SOUSTELLE Michel
Handbook of Heterogenous Kinetics